BrightRED Revision

Higher BIOLOGY

David Lloyd and Geoff Morgan

First published in 2010 by:

Bright Red Publishing Ltd
6 Stafford Street
Edinburgh
EH3 7AU

MIX
Paper from responsible sources
FSC® C013254

Reprinted (with corrections) in 2011

A CIP record for this book is available from the British Library

ISBN 978-1-906736-21-7

With thanks to Ken Vail Graphic Design, Cambridge (layout), Anna Clark (editorial)
Cover design by Caleb Rutherford – eidetic
Illustrations by Beehive Illustration (Mark Turner) and Ken Vail Graphic Design

Acknowledgements
Every effort has been made to seek all copyright-holders. If any have been overlooked, then Bright Red Publishing will be delighted to make the necessary arrangements.

Bright Red Publishing are grateful for the use of the following:

Photo © University of Pittsburgh at Bradford/Public Domain (p 3); Photo © Daderot/CC-BY-SA* (p 11); Photo © Alcibiades/Public Domain (p 11); Photo © Jonas Schenk/Public Domain (p 11); Photo © Richard Wheeler/CC-BY-SA* (p 19); Photo © Roy van Heesbeen/Public Domain (p 25); Photo © Geoff Morgan (p 31); Photo © Mnolf/CC-BY-SA* (p 34); Photo © Paphrag/Public Domain (p 35); Image taken from Bolzer et al., (2005) Three-Dimensional Maps of All Chromosomes in Human Male Fibroblast Nuclei and Prometaphase Rosettes. Public Library of Science Journal: Biology 3(5): e157 DOI: 10.1371/journal.pbio.0030157, Figure 7a © Public Library (p 38); Photo © EMW/CC-BY-SA* (p 38); Photo © United States Department of Homeland Security/Public Domain (p 43); Photo © PaleWhaleGail/CC-BY-SA* (p 43); Photo © Geoff Morgan (p 44); Photo © Seb951/CC-BY-SA* (p 45); Photo © Orchi/CC-BY-SA* (p 47); Photo © Geoff Morgan (p 50); Photo © Patrioter 6/CC-BY-SA* (p 52); Photo © Ninjatacoshell/CC-BY-SA* (p 52); Photo © David Midgley/CC-BY-SA* (p 53); Photo © Geoff Morgan (p 54); 2 photos © Geoff Morgan (p 55); 3 photos © Geoff Morgan (p 56); 2 photos © Richard Bartz and Wolfgang Hägele/CC-BY-SA* (p 57); Photo © John Desjarlais/CC-BY-SA* (p 57); Photo © Richiebits/Public Domain (p 57); Photo © Piccollo Namek/CC-BY-SA* (p 57); Photo © Svdmolen/CC-BY-SA* (p 57); 2 photos © Geoff Morgan (p 58); Photo © Geoff Morgan (p 60); 2 photos © Geoff Morgan (p 61); 2 photos © Geoff Morgan (p 63); Photo © David Lloyd (p 66); 4 photos © David Lloyd (p 67); 4 photos © Geoff Morgan (p 68); Photo © Geoff Morgan (p 69); Photo © Mike Baird/CC-BY-SA* (p 71); Photo © Geoff Morgan (p 74); Photo © Geoff Morgan (p 75); Photo © NOAA/Public Domain (p 75); 2 photos © Microrao/Public Domain (p 75); Photo © USDE/Public Domain (p 75); Photo © Mystrica Ltd (p 76); Photo © Sherry Crawford (p 78); Photo © Matthiaskabel/CC-BY-SA* (p 78); Photo © Estormiz/Public Domain (p 78); Photo © Karel Schmeidberger/CC-BY-SA* (p 80); Photo © Grzegorz Polak/CC-BY-SA* (p 80); Photo © Mnolf/CC-BY-SA* (p 80); Photo © National Cancer Institute/Public Domain (p 80); Photo © Val Vannet/CC-BY-SA* (p 81); Photo © Felagund/CC-BY-SA* (p 81); Photo © Manuguf/CC-BY-SA* (p 82); Photo © Zerohund/CC-BY-SA* (p 82); 2 photos © Geoff Morgan (p 86); Photo © Geoff Morgan (p 87); Photo © Geoff Morgan (p 88); Photo © Per H. Olsen/CC-BY-SA* (p 88); Photo © Geoff Morgan (p 89); Photo © Viridiflavus/CC-BY-SA* (p 89); Photo © Xenus/Public Domain (p 90); 2 photos © Geoff Morgan (p 90); Photo © Michael Gäbler/CC-BY-SA* (p 92); Photo © Andreas Trepte/CC-BY-SA* (p 92); 4 photos © Geoff Morgan (p 93); 2 photos © Geoff Morgan (p 94); 3 photos © Geoff Morgan (p 95); 3 photos © Geoff Morgan (p 96); 3 photos © Geoff Morgan (p 97); Photo © United States Department of Health and Human Services/Public Domain (p 102); Figures taken from 'Causes of death' spreadsheet, table 'Age-std death rates', Column G Cardiovascular diseases (GBD code W104) http://www.who.int/whosis/en/ Reproduced by permission of WHO (p 103); Figure from p 1143 of C.J.K. Henry, "Basal metabolic rate studies in humans: measurement and development of new equations", Public Health Nutrition, Volume 8(7a): pp 1133-1152, (2005) (p 104); 2 photos © Alan Boyde (a.boyde@qmul.ac.uk) 2010 (p 108); 4 photos © Geoff Morgan (p 110); Photo © Geoff Morgan (p 111); 2 photos © Geoff Morgan (p 113); Photo © Geoff Morgan [photograph of pp. 268-269 of the book Colour Identification Guide to Moths of the British Isles by B. Skinner, 3rd ed., (2009 ISBN 978-87-88757-90-3), published by Apollo Books www.apollobooks.com] (p 115).

* Licensed under the Creative Commons Attribution-Share Alike 3.0 Generic license. Details can be viewed at: http://en.wikipedia.org/wiki/CC-BY-SA-3.0
** Licensed under the Creative Commons Attribution 2.5 Generic license. Details can be viewed at: http://creativecommons.org/licenses/by/2.5/

Printed and bound in the UK by Martins the Printers

Both authors wish to thank their families for their patience, tolerance and support.

David Lloyd thanks the Principal and his colleagues at Stewart's Melville College for their support.

Geoff Morgan thanks the Principal and his colleagues at George Watson's College for their support and the opportunity of a sabbatical term when many of the photographs for this book were taken.

Thanks also to all those who have helped scrutinise and produce this book. Any errors that remain are, of course, ours alone.

Bright Red Publishing would also like to thank the Scottish Qualifications Authority for use of Past Exam Questions. Answers do not emanate from the SQA.

CONTENTS

ADVANCED HIGHER BIOLOGY

AN OVERVIEW

USING THIS BOOK FOR REVISION

In this book we have tried to present the content of the Advanced Higher Biology course in a concise form which will help you to understand and to learn. It is, however, only an **aid to your revision**. Everyone finds their own way of revising effectively but it has to be an **active process**! Your brain needs to be doing some serious work; this means that you have to take in information, think about how you can understand it for yourself, and then write it out again using your own words or pictures. Finally, you must keep **returning** to your revision notes to reinforce your learning.

Revising is an active process.

This book has features to help your understanding and your learning:

- Some words or phrases are in **bold** to emphasise their importance. These are key terms and you should be able to explain what they mean, or be able to use them to explain other ideas. We have also tried to show how aspects of the course connect together by indicating other pages that are relevant.
- **Don't forget** items try to point out common errors so that you can avoid them. We have also included some ideas that may help you to remember key facts.
- We have included some **internet links** in blue boxes, like the one below. These have addresses for websites that will help to reinforce the ideas or show how they are relevant in real situations.
- **Let's think about this** is where we have tried to take the content of a particular page further. We hope that you find these interesting and that they also aid understanding.

We have also included some **practice questions** (see pages 116–121) to get you started. But to get some really good preparation for the actual exam you should work through the recent past papers.

Some useful tips for revision can be found at http://www.open.ac.uk/skillsforstudy/revision-techniques.php

THE COURSE

Mandatory units

The Advanced Higher Biology course has three **mandatory** units that you must study. These are shown in the table below with the number of hours that the SQA suggests they each will take to study.

Unit title	Unit size
Cell and Molecular Biology	Full (40 hours)
Environmental Biology	Full (40 hours)
Investigation	Half (20 hours)

Option units

There are three units which are **options** and you will study only one of them. The table below shows these, again with their suggested study times.

Unit title	Unit size
Biotechnology	Half (20 hours)
Animal Behaviour	Half (20 hours)
Physiology, Health and Exercise	Half (20 hours)

DON'T FORGET

You will study the two full units, the Investigation half-unit and one option half-unit.

Go to http://www.sqa.org.uk/sqa/2571.html to get the Advanced Higher Biology Arrangements, commonly called the syllabus. The content columns and notes columns explain the coverage of each unit.

UNIT ASSESSMENTS

Each unit has a NAB assessment, of course. The taught units have a test of short answer questions, with about 70% of the marks testing your knowledge and understanding (KU) and 30% testing your problem solving (PS).

Unit title	Test total	Pass mark
Cell and Molecular Biology	40	26
Environmental Biology	40	26
Option half-units	20	13

PROBLEM SOLVING

selecting information

presenting information

processing information

making predictions

drawing conclusions

evaluating experimental procedures

planning experimental procedures

The Investigation half-unit also has a unit assessment which takes the form of a Laboratory Notebook (see page 114).

Problem solving skills are assessed. More information about these can be found on page 56 of the Arrangements documents – see the internet link at the bottom of page 2.

DON'T FORGET

No marks from the unit assessments go towards the final mark. But you must pass each one to be able to get a course award on your final certificate.

HOW THE COURSE IS GRADED

Your final grade is based on 125 marks. The written exam has 100 marks and the Investigation report has 25 marks. Guidance for writing your Investigation report is on pages 114–115.

The written exam

This takes 2½ hours and is out of 100 marks. About 70 marks test your knowledge and understanding (KU), and 30 marks test your problem solving (PS). There are three sections in the exam: Sections A and B test the two mandatory full units; Section C tests the option units. The table below shows the structure of the exam.

Practical work should give you a deeper understanding and help your attainment in the final exam.

	Section A	Section B	Section C
Marks	25 marks	55 marks	20 marks
Question types	25 multiple choice questions	• data question (about 15 marks) • short answer questions (about 20 marks) • mini essay (about 5 marks) • one from a choice of two essays (15 marks)	• short answer questions (about 15 marks) • mini essay (about 5 marks)
KU and PS allocation	KU = about 15 PS = about 10	KU = about 40 PS = about 15	KU = about 15 PS = about 5
Content	Cell and Molecular Biology, and Environmental Biology		option units

DON'T FORGET

Section C has questions on all three option units. Of course, you only answer questions on the one that you have studied.

LET'S THINK ABOUT THIS

Broaden your interest in biology. There are lots of very readable popular science paperbacks about many different fields of biology – look in your library, bookshop or on line. You may be able to subscribe to *Biological Sciences Review* which is written for S6 students. *NewScientist* regularly has interesting articles about cutting-edge biology.

PROKARYOTES AND EUKARYOTES

CELLS ARE THE BASIS OF LIFE

All organisms are made from one or more cells. All the life processes of metabolism and heredity are carried out within these cells. They are the smallest unit of life and they are all direct descendants of the first cells that evolved on the early Earth. Although these first cells arose spontaneously from the conditions of the early Earth, modern cells can only be made by the division of a previously existing cell.

Organisms can be divided into two major groups, depending on the structure of their cells. These groups are the **eukaryotes** (the protists, fungi, animals and plants) and the **prokaryotes** (the bacteria and archaea). Eukaryote means 'true nucleus' while prokaryote means 'before the nucleus'. However, this may be wrong in terms of evolution; prokaryotes may well have arisen from simplified versions of early eukaryotes!

PROKARYOTES

The bacterial cell shown below is an example of a prokaryote.

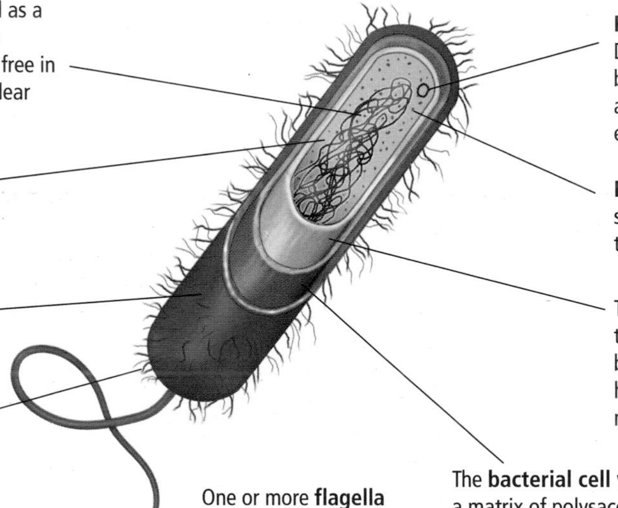

A prokaryotic cell (typical size 2–10 µm)

DON'T FORGET

The genetic material of prokaryotes is circular DNA. This is often referred to as a 'bacterial chromosome'.

EUKARYOTES

Structures in eukaryotic cells

Cell structure in eukaryotic cells is much more complex than in prokaryotic cells.

- Eukaryotic cells have **a nucleus** with **nuclear membrane** as a boundary. This boundary is a double layer of membranes with pores to let very large molecules pass through.
- The genetic material is **linear** DNA organised as proper **chromosomes**. The DNA is wrapped around **histone proteins** to form **nucleosomes** as shown in the diagram opposite.
- The **plasma membrane** has many functions (see page 22). The **cytosol** is the liquid part of the cytoplasm. It is organised by **endomembranes** which compartmentalise different parts of the metabolism.
- The endomembranes form the **endoplasmic reticulum** (ER) and the **Golgi apparatus**.
- **Ribosomes** are for protein synthesis and they are larger than those found in prokaryotes.

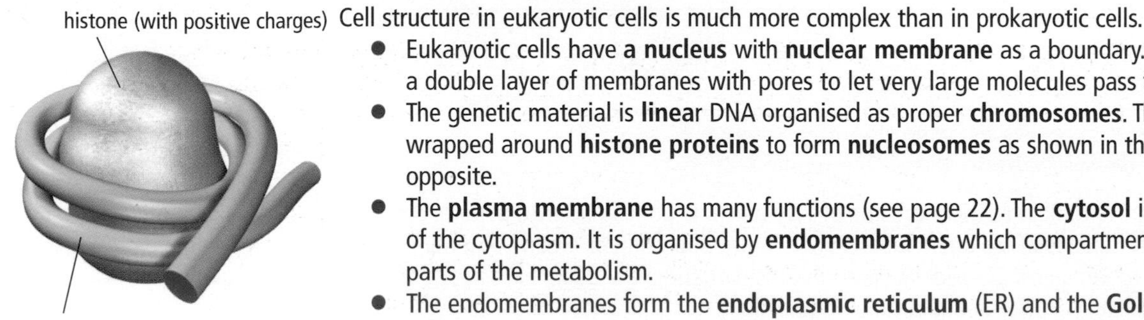

A nucleosome

contd

EUKARYOTES contd

- **Mitochondria** are present for aerobic respiration. These have a double-layer of membranes with the inner layer folded to form cristae.
- **Microbodies** only have a single membrane around them and are produced by budding from the ER or the Golgi apparatus. **Lysosomes**, which contain enzymes to digest and recycle parts of the cell, are examples.
- The cytosol has a **cytoskeleton**. See page 25 for more about this important part of the cell.

Animal cells

Animal cells have some structures that are unique to them.

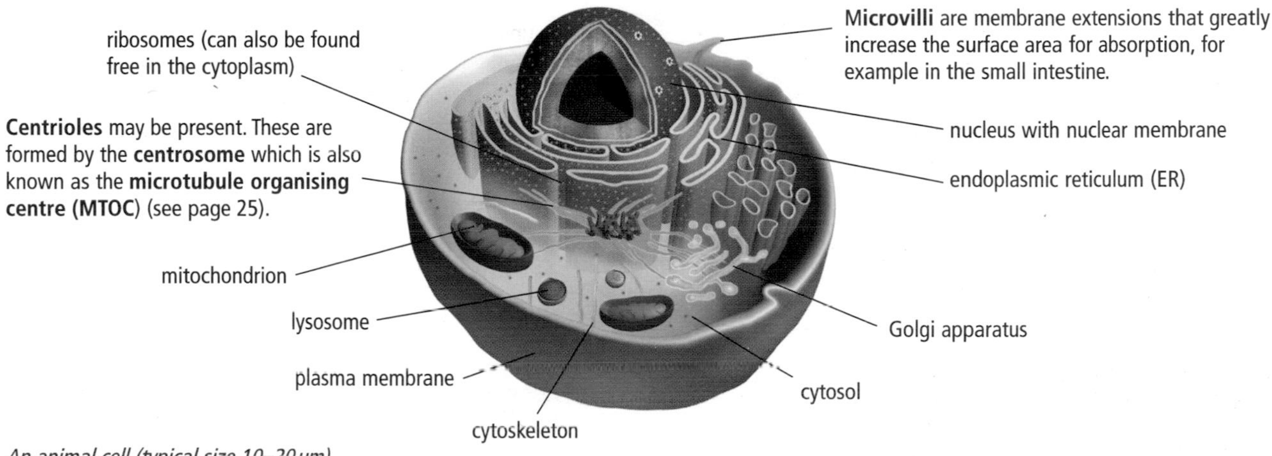

An animal cell (typical size 10–20 µm)

Plant cells

Plant cells also have some structures that are unique to them. Plants cells have more than one MTOC and these are distributed close to the nucleus of the plant cell.

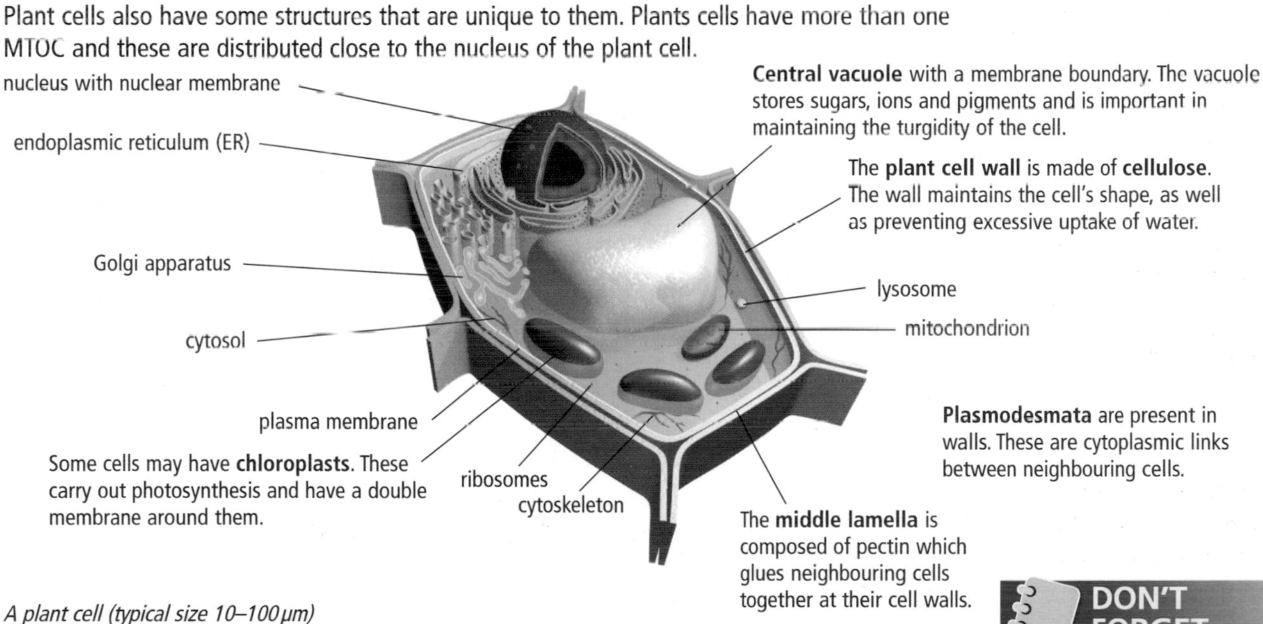

A plant cell (typical size 10–100 µm)

DON'T FORGET

As well as special features of their DNA and nucleus, eukaryotes have a cytoskeleton and membrane compartments in their cytoplasm.

Go to http://www.biology.arizona.edu/CELL_BIO/tutorials/pev/main.html and follow the links to 'Organization', 'Prokaryotes' and 'Eukaryotes'.

LET'S THINK ABOUT THIS

The *endosymbiont theory* suggests that mitochondria and chloroplasts are the descendants of prokaryotes that were absorbed into early eukaryotic cells. What evidence is there for this idea? Mitochondria and chloroplasts do have some prokaryotic features: they have a circle of DNA with their own genes; and they have ribosomes that are smaller than in the rest of the eukaryotic cell.

CELL CYCLE

THE SEQUENCE OF EVENTS IN THE CELL CYCLE

Some eukaryotic cells are able to reproduce themselves by following a sequence of events in which their contents are duplicated and divided in two. The repeating sequence is referred to as the cell cycle. The cell contents are built up during **interphase**, which has three phases called G1, S, and G2. Interphase is followed by the M phase, which is composed of mitosis (the division of the nucleus) and cytokinesis (the division of the cytoplasm to form two cells). So, the whole cell cycle is **G1 → S → G2 → M.**

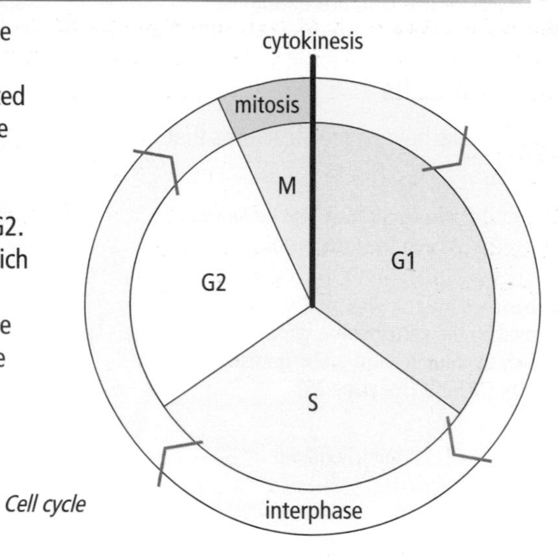

Cell cycle

Interphase – G1, S and G2 phases

In actively dividing tissue, cells in interphase appear to be doing nothing, but actually this is a very **active period** of growth and metabolism.

- **G1** is the first **growth stage** – the cell makes new proteins and copies of the organelles.
- **S** phase is when **DNA replication** occurs.
- **G2** is second period of **cell growth** – again, the cell makes more proteins and copies the organelles in preparation for mitosis.

Mitosis and cytokinesis – the M phase

Mitosis is a series of phases which flow from one to the next to divide the nucleus. **Spindle fibres** move the chromosomes during mitosis. These fibres are part of the cytoskeleton (see page 25).

Prophase: the chromosomes condense (coil up) and become visible as two joined chromatids. Spindle fibres attach to the chromosomes at centromeres. The nuclear membrane disintegrates.

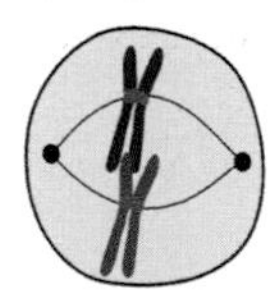

Metaphase: spindle fibres move the chromosomes so that they line up on the **metaphase plate at the equator** of the cell.

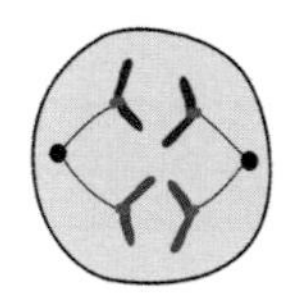

Anaphase: the spindle fibres pull the sister chromatids apart. Once they are separated, the chromatids are called chromosomes in their own right.

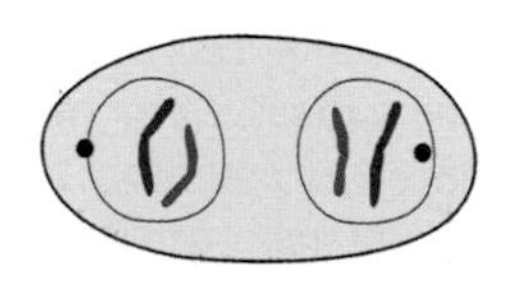

Telophase: the separated chromosomes are pulled by the spindle fibres to opposite poles to form **daughter nuclei**. The chromosomes start to uncoil and a nuclear membrane is made again.

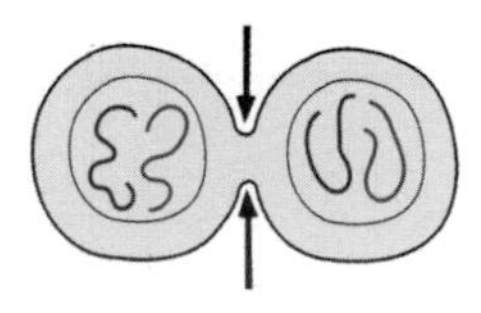

Cytokinesis: the membrane is pulled in by part of the cytoskeleton to divide the cytoplasm to form two daughter cells. Plant cells have to form a middle lamella and cell wall before the membrane is made.

Mitosis and cytokinesis

contd

THE SEQUENCE OF EVENTS IN THE CELL CYCLE contd

Mitotic index

The cell cycle can vary in duration depending on the tissue. A cell in the growing points of a mature plant may stay in G1 for weeks, whereas embryonic sea urchin cells can go through a complete cycle in two hours. When a sample of cells is viewed, the **mitotic index** can be found; this is the **percentage of cells undergoing mitosis** in the sample. If a tissue sample has an unusually high mitotic index, this may indicate a developing tumour.

DON'T FORGET

When you are asked about the cell cycle, it is important that you know the details of the interphase stages (G1, S and G2) as well as the stages of the M phase.

CONTROL OF CELL CYCLE

The cell cycle has many complex events that all have to work perfectly to produce new daughter cells. There are three major checkpoints which act as gatekeepers that only allow progress to the next stage if everything is ready and correct.

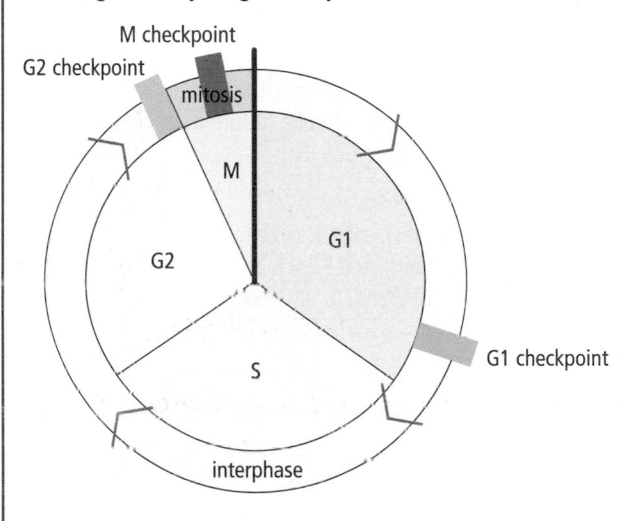

Cell cycle with checkpoints

The **G1 checkpoint** is near the end of G1. Here the **cell size is monitored.** There has to be sufficient cell mass to form two daughter cells. This checkpoint controls entry to the S phase.

The **G2 checkpoint** is at the end of G2. This checks the **success of DNA replication** to make sure each daughter cell can receive a complete copy of DNA. This checkpoint controls entry to mitosis using a protein complex called **mitosis-promoting factor** (MPF). As the name implies, mitosis will only be triggered if sufficient MPF is present.

The **M checkpoint** is during metaphase. This monitors the **chromosome alignment** to ensure each daughter cell receives one chromatid from each chromosome. This checkpoint controls the entry to anaphase. Thus, it triggers the **exit from mitosis and the start of cytokinesis**.

DON'T FORGET

Remember the exact timings of the checkpoints. The G1 checkpoint is near the end of the G1 phase. The G2 checkpoint is at the end of G2 phase. The M checkpoint is during metaphase.

Do you know enough about the cell cycle to operate the machinery yourself in an interactive game? Try at http://nobelprize.org/educational_games/medicine/2001/

LET'S THINK ABOUT THIS

One of the genes involved in the G1 checkpoint is called *p53*. This codes for a protein which checks that the replicated DNA strands are undamaged. If damage is detected, then *p53* protein stops the cycle and repair enzymes are brought in. If the damage is repaired, then the cycle can continue. If not, then *p53* instructs the cell to kill itself. The *p53* protein has been found to be missing or faulty in most cancerous cells – one reason why these cells can divide without being halted at the G1 checkpoint (see page 9).

NORMAL DIFFERENTIATION AND ABNORMAL DIVISION

NORMAL DIFFERENTIATION

Multi-cellular organisms have **cells that are specialised** for particular functions. These cells are described as **differentiated** – they have different structures and different functions; muscle cells are specialised to contract and nerve cells pass signals. Differentiated cells which have the same function are grouped together into **tissues**, so we have muscle tissue and nerve tissue for example. Tissues are grouped together into **organs** which combine the functions of the tissues; the stomach has muscles to churn the food and nerves to detect when the stomach is stretched and full.

The growing points of plants have regions of cells that are capable of cell division. These growing points are called **meristems**. Meristem cells in plants are **totipotent** – they can differentiate into any other type of cell. Mammals have regions of cells (adult stem cells) that are specialised and so can only produce a restricted range of cell types. For example, bone stem cells can only produce new bone cells.

DON'T FORGET

All cells in an organism have the same set of genes. It is the changes in gene expression that produce differentiated cells.

With a few exceptions (such as red blood cells and gametes), all the cells in an organism have the same set of genes. So how do the cells become differentiated? Each cell type has a different combination of **genes being expressed**; some genes have been switched on and others have been switched off. This effect is permanent, explaining why **specialised cells can no longer differentiate** into other cells.

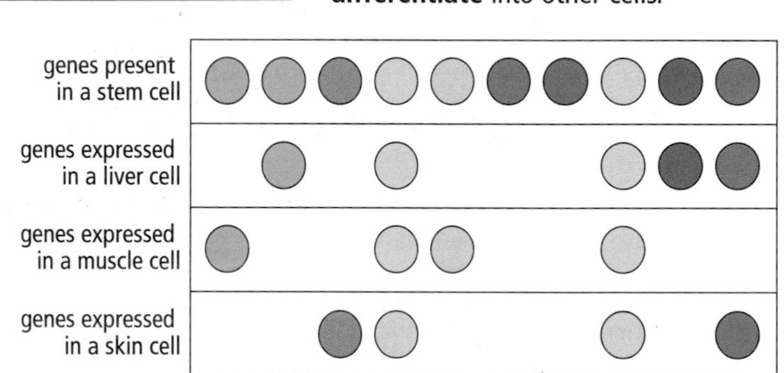

All cells in an organism contain the same genes. In a differentiated cell only some of these genes are expressed.

In animals, **embryonic stem cells** have the ability to **differentiate into any cell type** but this property is lost as the organism develops. Adult stem cells have become specialised and will not normally produce other cells. It is possible that embryonic stem cells could be used to replace damaged cells in adults, but there are ethical objections to this as the stem cells have to be gained from embryos. However, it is now possible to treat adult stem cells so that they behave like embryonic stem cells, removing some potential ethical problems.

CONTROL OF GENE EXPRESSION

In eukaryotes, gene expression is likely to be controlled by a variety of mechanisms. These may include changes to the histone protein of the nucleosome (see page 4) or methylation of cytosine nucleotides.

Gene switching in prokaryotes is much better understood. The first example to be worked out was the ***lac* operon** in *Escherichia coli*. Jacob and Monod put forward their hypothesis in 1961.

The system has:

- a **regulator gene** which codes for a repressor molecule
- the **repressor molecule** which binds to the operator or the inducer
- a **structural gene** that makes β-galactosidase, an enzyme which breaks down lactose
- an **operator** consisting of a length of DNA 'upstream' of the structural gene
- an **inducer molecule** (lactose) which binds to the repressor molecule.

When no lactose is present the system works like this. The regulator gene makes the repressor molecule. This attaches to the operator so the transcription enzyme cannot read the structural gene. No β-galactosidase is made and the bacterium does not waste energy making an enzyme it does not need.

contd

CONTROL OF GENE EXPRESSION contd

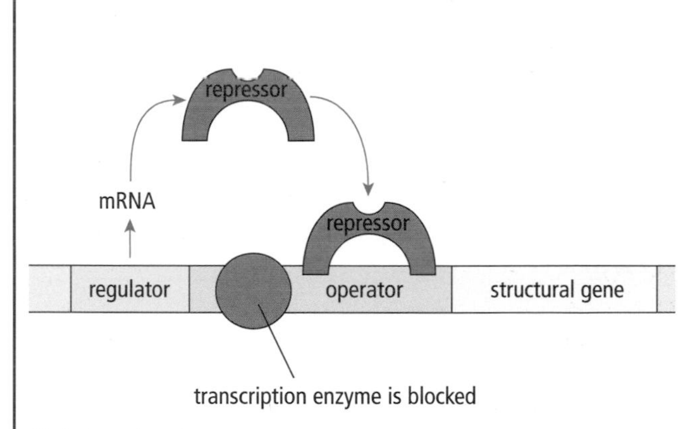

No lactose in cell

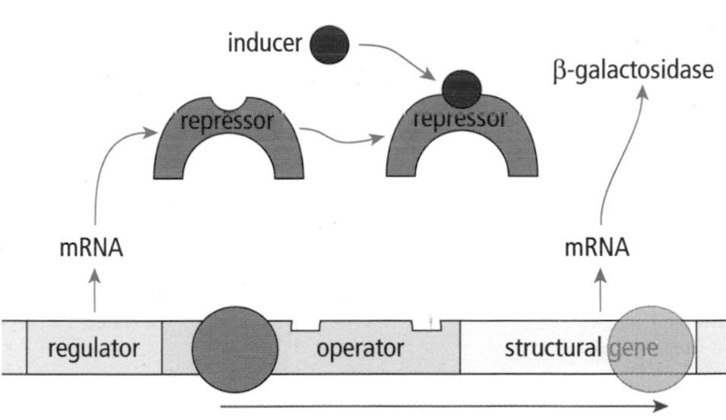

Lactose is present

When lactose is present, it acts as the inducer to cause the production of β-galactosidase. The regulator still makes the repressor molecule but now the inducer attaches to it. This leaves the operator open so the transcription enzyme can read the structural gene; β-galactosidase is produced and lactose can be broken down. (To learn more about this system go to page 77.)

DON'T FORGET

Try to understand the workings of the *lac* operon. The names of the key structures really do tell you what they do!

ABNORMAL CELL DIVISION

Proliferation genes

These genes code for proteins which **promote cell division** only when an external signal is received, for example when a wound needs to be repaired. They are also known as **proto-oncogenes** because, when they mutate, they form **oncogenes** which are found in many cancers. These mutations give proteins with abnormal shapes which stimulate **excessive cell division** and **tumour formation** even in the absence of any external signal. Because a mutation only has to happen in one gene of the pair, oncogenes can be thought of as **dominant** alleles.

A mutation to a proliferation gene is like a car's accelerator being jammed ***on****!*

Anti-proliferation genes

These genes code for proteins which restrict cell division by operating at the cell cycle checkpoints. They are also known as **tumour suppressor genes** because they normally prevent excessive cell division. One functioning gene can still produce the protein to inhibit the cell cycle, so **both copies** of the gene must mutate before the control of the cell cycle is lost and a tumour starts to form. For this reason, the cancer-causing mutations to anti-proliferation genes can be thought of as **recessive**.

Anti-proliferation genes are like brakes on a bike going downhill. ***Both*** *sets need to stop working before control is lost.*

Type 'lac operon' into YouTube for a good 3.24 minute animation of this system.

DON'T FORGET

For cancers to develop, proliferation genes only need one copy to mutate, while both copies of the anti-proliferation genes must mutate.

LET'S THINK ABOUT THIS

The Jacob–Monod hypothesis of the regulation of the *lac* operon is an example of a negative feedback system. The gene is switched on when it is needed and switched off when it is not required – this conserves energy and resources. Negative feedback systems function at all levels of biology, from genes and metabolism, through to physiology and ecosystems.

CELL AND TISSUE CULTURE

PRINCIPLES OF CULTURING CELLS

Cell and tissue culture is the growth of cells, tissues or organs in artificial media in a laboratory. Culturing can produce many genetically identical **clones** of an initial cell sample. The term **cell culture** is often used to describe the growth of single-celled organisms or individual cell types, whereas **tissue culture** is used more widely.

General culture requirements

Some cell types can be grown in **suspension** using liquid medium. Other types can be grown on, or within, a solid medium (**substrate**) such as **agar jelly**. To ensure rapid growth of the chosen cells, optimum conditions are provided in terms of nutrients, pH, and gases. Anaerobic organisms can be cultured in non-aerated suspension or within agar.

Aseptic techniques

Contamination is the enemy of cell or tissue culture. Aseptic techniques, such as the use of **sterile materials** and the treatment of source tissue, are vital to prevent inoculation with unwanted cells or spores. Bacterial or fungal contamination will rapidly outcompete and **spoil** a culture of slower-growing plant or animal cells.

BACTERIAL AND FUNGAL CELL CULTURES

Bacteria and fungi often have simple culture requirements. Many strains are easily cultured in suspension in fermenters. In schools, bacteria and fungi are often grown in Petri dishes on **nutrient agar** (agar with peptides and beef extract added) or **malt agar** (agar with malt extract added). Plates are incubated at temperatures other than human body temperature to reduce the chance of culturing pathogens. (For more detailed information on the culture of organisms, see page 76.)

MAMMALIAN CELL CULTURE

Specific culture requirements

To undergo rapid growth and cell division in culture, most mammalian cells require the addition of a complex medium containing chemical **growth factors**. Growth factors can be provided by the addition of an **animal serum**, such as foetal bovine serum (FBS). FBS is a mixture containing growth factors, proteins, salts, vitamins and glucose. **Antibiotics** are also added to minimise the chances of spoilage by microorganisms.

DON'T FORGET

Memorise the sequence of events that occurs during the subculture of anchorage-dependent cells: adhering, spreading, dividing, formation of confluent monolayer.

Subculturing

Mammalian cells are usually cultured in a flask. Cells for culture are detached from source tissue using **proteolytic enzymes** such as trypsin. When the cells are added to the flask they **adhere** to the surface of the agar, then flatten or **spread out**, and then start to **divide**. The cells form a **monolayer**, a layer one cell thick. They stop dividing when they are **confluent**; that is, when they have 'flowed together' to form a complete covering of the surface. Cells soon use up the nutrients in the medium, so, to keep the cloned cell line alive, some must be **subcultured** into a fresh culture flask. During this process, cells have to be detached from the monolayer, again using proteolytic enzymes.

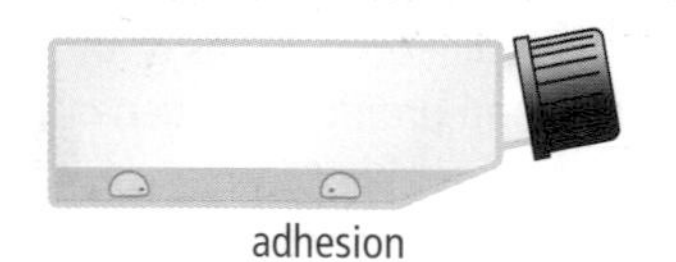

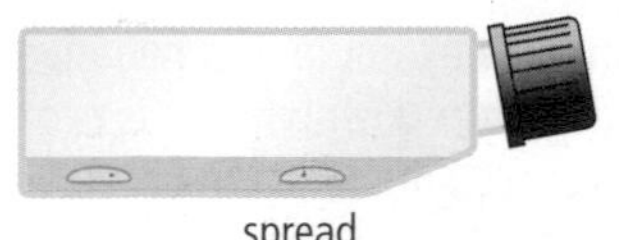

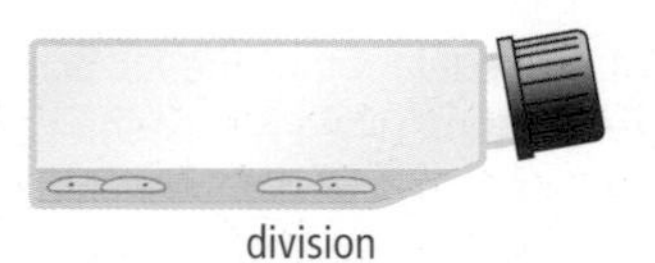

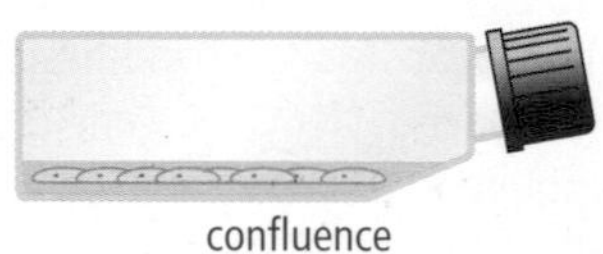

Mammalian cells becoming established in a tissue culture

contd

MAMMALIAN CELL CULTURE contd

Selection of cell lines

Normal animal cells tend to die after a finite number of divisions (about 60 – the Hayflick limit). This reduces the length of time a culture can be maintained. However, this limit does not exist in **immortal cell lines**, for example those derived from cancer cells or stem cells which can be subcultured indefinitely. **Stem cells** are much sought after for tissue culture as they can be **pluripotent** (able to differentiate into many cell types) or even **totipotent** (able to differentiate into all cell types).

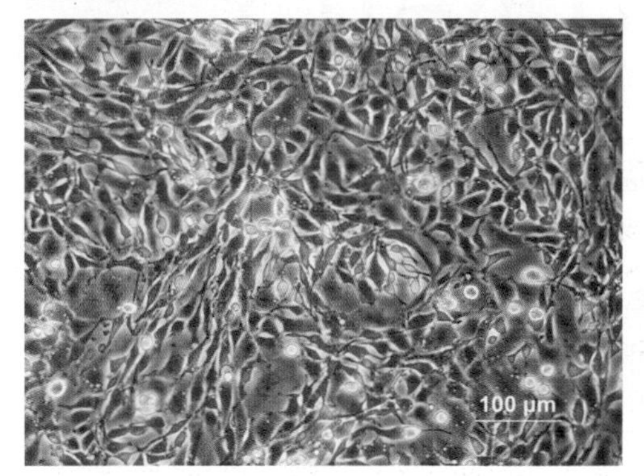

A monolayer of anchorage-dependent mammalian cells

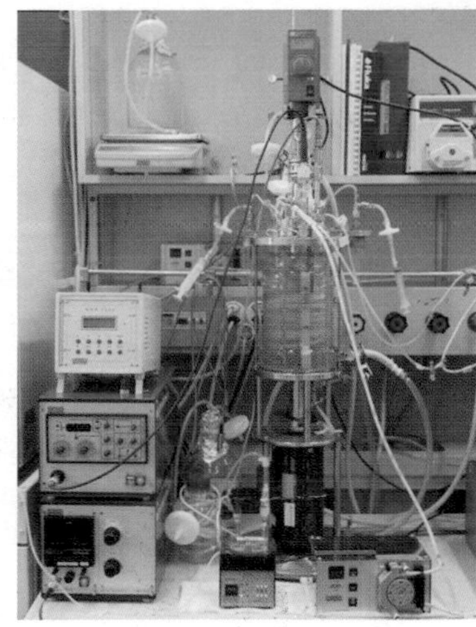

A laboratory fermenter is set up to grow a suspension culture of anchorage-independent mammalian cells under carefully controlled conditions.

PLANT TISSUE CULTURE

Specific culture requirements

To make a growth medium suitable for plant tissue culture, Murashige and Skoog (M&S) salts are often used. These salts contain an appropriate balance of **macronutrients** (for example N, P, K, Mg), **micronutrients** (such as Zn, Na, Cu), **carbon** sources (sugars) and vitamins. **Growth regulators**, such as cytokinins and auxins, can also be added to stimulate differentiation.

Plant tissue culture

Selection of plant cells

Plant cells are **totipotent** (that is they have 'total potential') and so are able to differentiate into all the cell types required to form a whole new organism. The two main methods of plant tissue culture use either **explants** or **protoplasts**.

Explants are small pieces of plant tissue that are placed on a solid medium to either promote shoot growth or callus growth. **Organ formation** is then stimulated by altering the ratio of **cytokinins** (to promote shoot growth) and **auxins** (to promote root growth).

Protoplasts are created by removing cell walls with the enzymes **pectinase** and **cellulase**. The resulting cells can be grown in liquid medium as a **cell suspension culture** and treated with growth regulators to induce **embryogenesis** to generate whole new embryonic plants. Protoplasts can also be encouraged to form a **callus**, as with explants. For protoplast fusion, see page 80.

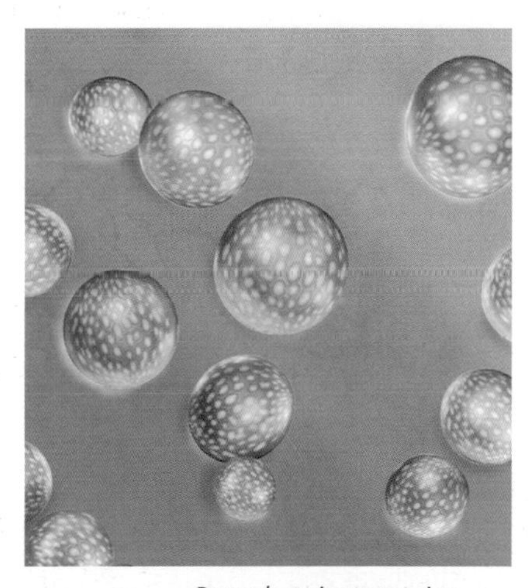

Protoplasts in suspension

Purpose of plant tissue culture

Cell suspension cultures allow **screening** for beneficial traits (such as heat or salinity tolerance) or the selection of **virus free cells** for cloning. **Micropropagation** – the generation of many cloned plantlets from one source plant – allows rapid upscaling to field trials. In addition, cell suspension cultures are used to harvest useful plant cell secretions.

Search for *Banana tissue culture in action* on YouTube for a clip about the impact tissue culture can have on developing communities.

DON'T FORGET

A suitable source of carbon for autotrophs can be CO_2 – as long as light is provided and the tissue being cultured is photosynthetic.

LET'S THINK ABOUT THIS

Could a doctor start a cell line using your cells without your permission? Do human cell lines have 46 chromosomes? Why do some cell lines become 'laboratory weeds'? Find out more about the *HeLa* cell line to answer these questions.

CARBOHYDRATES

GLUCOSE

Glucose is the **monosaccharide** which forms the building block for the **polysaccharides**.

The structure of glucose

The glucose monomer can exist in a **linear form** and so can be drawn out as a straight line of 6 carbons. Carbon 1 (or 1′) is shown at the top. However, the chain form is really a C-shape due to the 3-D angles of the chemical bonds. This brings the OH of 5′ near to the oxygen of 1′.

The linear and the **ring forms** are interchangeable and they **exist in an equilibrium** when in solution. But this equilibrium is heavily skewed to the ring form; for every one in linear form, there are 4000 of the ring form! The ring form is a perspective view of the molecule with 2′ and 3′ coming towards you. If you look closely at the ring form diagrams you can see that a glucose molecule can take on two slightly different structures. If the OH group of Carbon 1 is **below the ring** then it is **α-glucose**. If the OH group of Carbon 1 is **above the ring** then it is **β-glucose**.

actually folds into a 'C'

The linear form of glucose

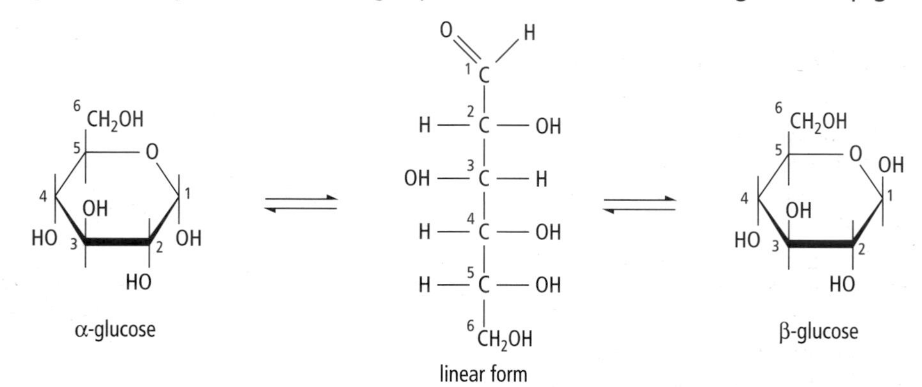

The equilibrium of glucose structures. (In the ring forms, corners represent carbon atoms and the hydrogens are not shown on the ends of bonds.)

DON'T FORGET

ABBA will help you remember the positions of the OH on Carbon 1. **A**lpha **b**elow, **b**eta **a**bove.

Linking the glucose monomers

Glucose monomers are linked together by enzyme-catalysed reactions. These enzymes cause a **condensation reaction** between the OH groups of Carbon 1 and Carbon 4 and so lead to the removal of a water molecule. The bond formed to link the glucoses is called a **glycosidic bond**.

Monomers of α-glucose are linked by an **α-1,4-glycosidic bond**. The bond between β-glucose monomers is a **β-1,4-glycosidic bond**.

DON'T FORGET

You should be able to draw glucose structures and use them to explain the condensation reaction.

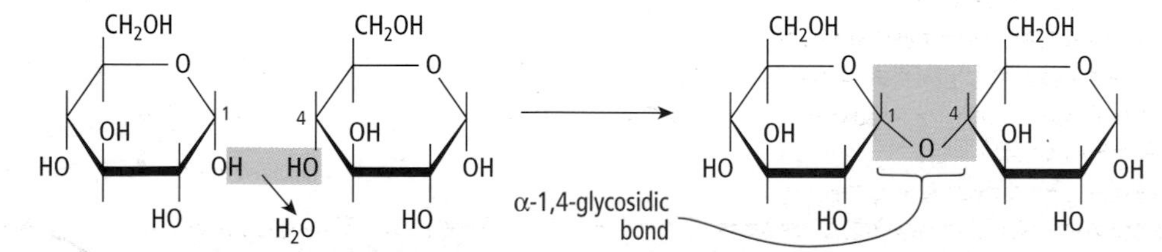

Forming an α-1,4-glycosidic bond between two α-glucose monomers. (Note that the single hydrogens are not shown.)

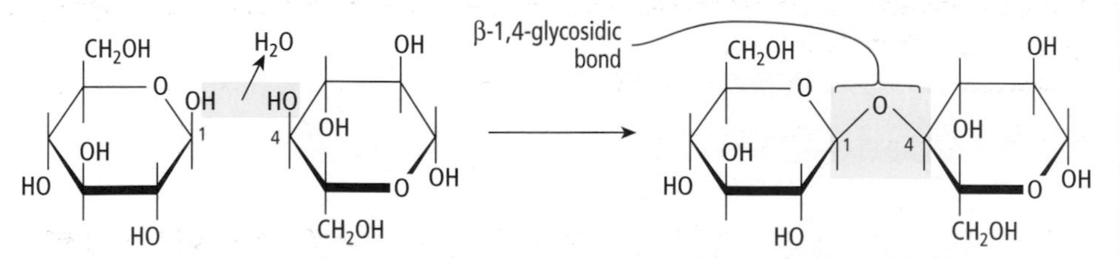

Forming a β-1,4-glycosidic bond between two β–glucose molecules. (Notice that the glucose monomer on the right is inverted.)

STORAGE POLYSACCHARIDES

Starch

The excess glucose made during photosynthesis can be built up to form starch in plant cells for **energy storage**. Starch is made from thousands of α-**glucose** monomers; there are two forms of starch.

α-1,4-glycosidic bond

Bonding in amylose, and the amylose helix

Amylose has many glucose monomers joined by α-**1,4-glycosidic bonds**. This forms a **long, linear molecule** with no side branches. Because of the bond angles, amylose is actually twisted and forms a long **unbranched helix**.

Amylopectin also has long chains of α-1,4-linked glucoses but it also has **side branches** every 25–30 units. These branches are linked to the main chains by α-**1,6-glycosidic bonds** which are formed between Carbon 1 and Carbon 6.

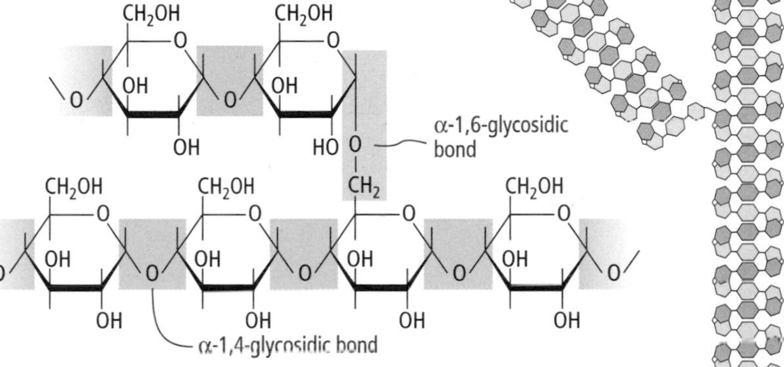

Bonding in amylopectin and the branched chain structure

Glycogen

Glycogen is also made from thousands of α-**glucose** monomers and is made in animal cells for **energy storage**. Glycogen is similar in structure to amylopectin with glucose monomers in long chains of α-1,4 linked glucoses with α-**1,6 side branches**. The branches are every 10–12 units and so there is a higher degree of branching in glycogen than in amylopectin. Glucose units are removed from the ends of the chain when required. So, since there are more ends in glycogen, glucose can be released quicker from glycogen molecules than from amylopectin molecules.

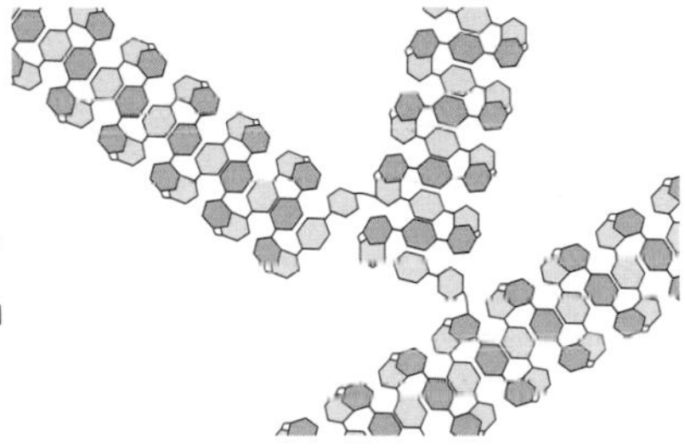

The branched chain structure of glycogen

Why do organisms build storage polysaccharides?

Why do organisms not just store glucose? A high concentration of glucose tends to draw water into a cell by osmosis, leading to problems for the cell. The huge starch and glycogen molecules are **insoluble**; they do not affect osmosis and so **osmotic problems are avoided**.

DON'T FORGET

Storage polysaccharides are made of α-glucose monomers.

CELLULOSE

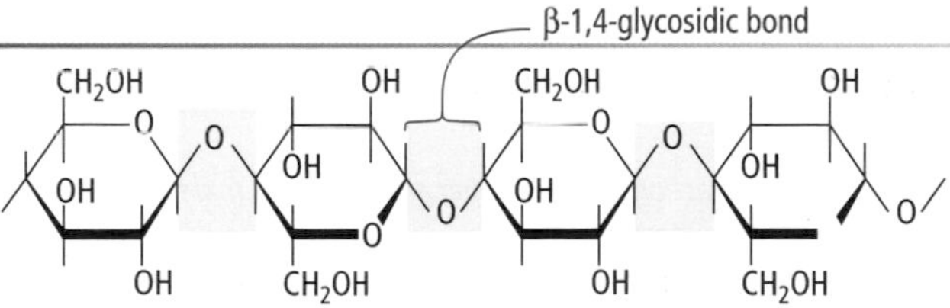

The bonding in cellulose

Cellulose is the **structural** carbohydrate providing support and rigidity in plant cell walls; it is made from thousands of **β-glucose** monomers joined by **β-1,4-glycosidic bonds**. Every alternate glucose unit is **inverted** producing a rigid **straight chain** with no branches. Because of the inverted glucose units, hydrogen bonds form between cellulose chains. Groups of about 40 cellulose chains are aligned and cross-linked to form fibrils. Many fibrils, arranged in layers, produce the cell wall.

Go to a search engine and type *Extra DNA causes Mendel's peas to pucker*. Select the item in findarticles.com to read about how an inability to make α-1,6-glycosidic bonds causes Mendel's wrinkled peas.

LET'S THINK ABOUT THIS

Mammals are unable to produce enzymes to digest the β-1,4-glycosidic bonds of cellulose. Ruminant animals (such as cattle) have symbiotic microbes in their guts that ferment the cellulose (see page 60). Unfortunately, the fermentation process also releases methane gas which contributes to the greenhouse effect (see page 72).

LIPIDS

TRIGLYCERIDES

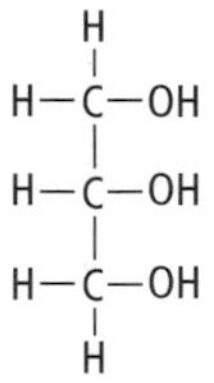

The structure of glycerol and fatty acids

Triglycerides are composed of **three fatty acids** linked to a **glycerol** molecule. Glycerol has three carbons, each with an **OH group**. Fatty acids have a **carboxylic acid** group (COOH) at one end and a **long hydrocarbon tail** (composed of only carbon and hydrogen) at the other.

Saturated fatty acids have have no double bonds between the carbon atoms in their tails. This means that a saturated fatty acid has the maximum number of hydrogen atoms that can be attached – so it is 'saturated' with hydrogens. **Unsaturated fatty acids** have at least one double bond between the carbon atoms and so have fewer hydrogen atoms than the maximum possible.

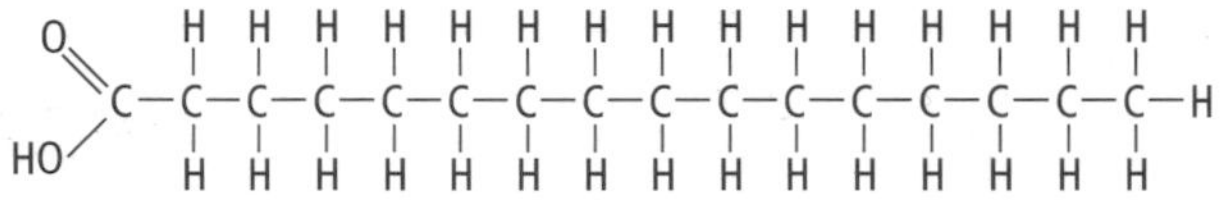

Glycerol

Saturated fatty acid

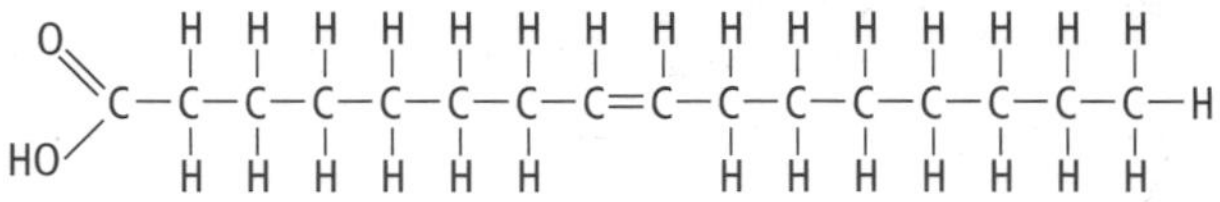

Unsaturated fatty acid

Linking glycerol and fatty acids

Glycerol and fatty acids are linked by enzyme-catalysed reactions. These are **condensation reactions** between the **OH of the acid** and the **OH of the glycerol**, resulting in the removal of a water molecule. The bond that is formed between the glycerol and the fatty acid is an **ester bond**.

The types of fatty acids found in a triglyceride will determine its physical properties. Triglycerides that are mainly composed of **saturated** fatty acids are **solid** at room temperature; they are fats. Oils are mainly composed of **unsaturated** fatty acids and are **liquid** at room temperature. The double bonds between carbons stop unsaturated fatty acids being packed close together and so the oil remains liquid at room temperature.

DON'T FORGET

The ester bond can remembered as a **COCO** bond. Look at the diagram to see how this works.

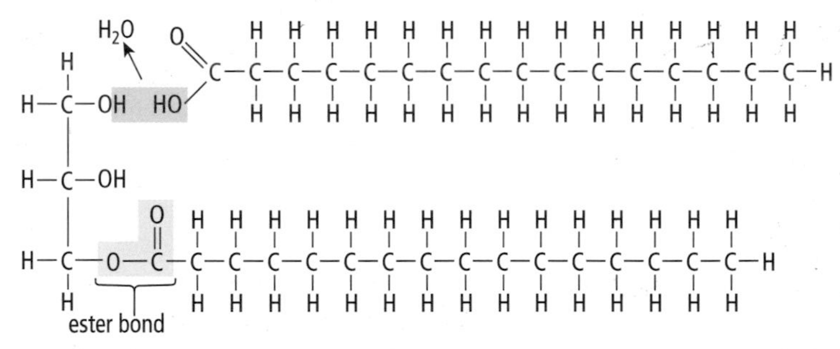

A condensation reaction (blue) forming an ester bond (pale green) to link a fatty acid to glycerol

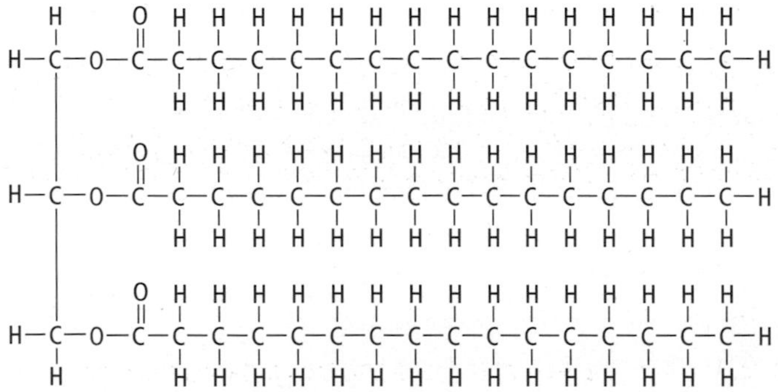

A triglyceride. Can you find the three ester bonds? Are the fatty acids saturated or unsaturated?

PHOSPHOLIPIDS

Phospholipids are very similar to triglycerides but they have a phosphate group in place of one of the fatty acids. This means that they are composed of **glycerol, two fatty acids** and a **phosphate**. The fatty acids and the phosphate are joined by **ester bonds** to the glycerol. Usually, one of the fatty acids is saturated and the other is unsaturated. Other charged groups, such as **choline**, may be attached to the phosphate.

The 'head' of a phospholipid is formed by the phosphate and the choline groups. Both of these groups are electrically charged which makes them hydrophilic – they are attracted to water. The 'tail' of a phospholipid is formed by the fatty acids, one of which is unsatutared. The fatty acids are uncharged and non-polar, which makes them hydrophobic – they are repelled by water.

The phospholipid bilayer in membranes forms because the hydrophilic heads stay in contact with aqueous solutions (extracellular fluids or the cytosol) and the hydrophobic tails stay in the middle of the bilayer, away from the water (see page 22). The unsaturated fatty acids in the tails keep the membrane fluid by preventing the phospholipids packing closely together.

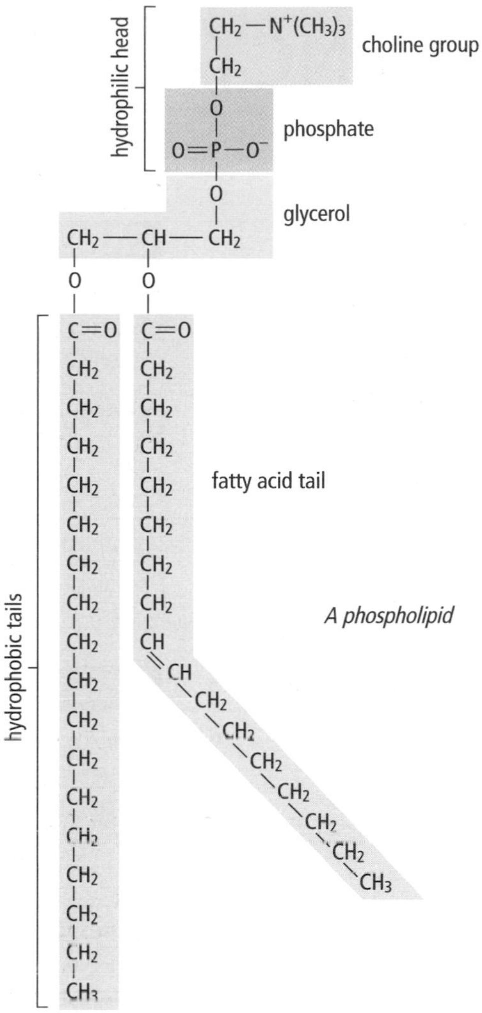

A phospholipid

STEROIDS

Steroids have a completely different structure to the other lipids. They are based on a **four-ring structure** of 17 carbons. Varying side chains attached to this structure give different steroids. Many steroids play a role in cell signalling as hormones; they are **hydrophobic** molecules so they are able to diffuse across membranes and bind to receptors inside cells (see page 31).

The two examples which we look at in Advanced Higher Biology are **cholesterol** and **testosterone**. The side chains vary in these molecules. Notice that:

- Cholesterol has an OH at the bottom left and C_8H_{17} at the top right.
- Testosterone has a double-bonded oxygen (C=O) at the bottom left and an OH at the top right.

Cholesterol

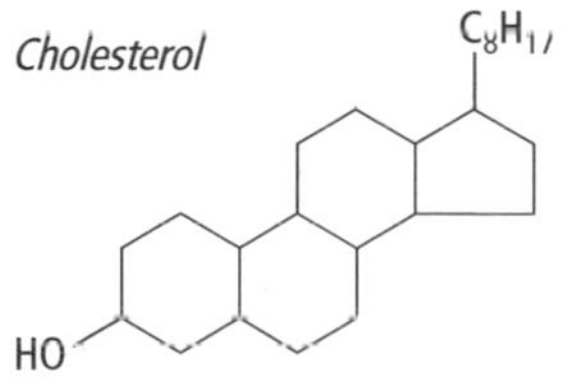

Testosterone

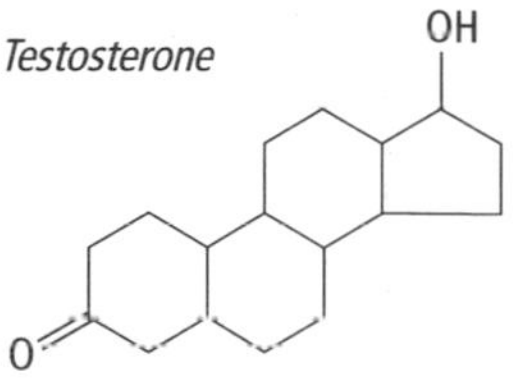

Testosterone is important in muscle development and in male puberty. Cholesterol fits into the hydrophobic part of the phospholipid bilayer and reduces phospholipid movement in the membrane. It also helps to prevent solidification of the membrane at low temperatures (see page 31).

Anabolic steroids are used by some people to help build up muscle or allow them to train harder. But there are some drastic side effects. Read about these at http://www.medicinenet.com/script/main/art.asp?articlekey=52945

DON'T FORGET

You should practise drawing out the four-ring structure of the steroids. You also need to remember the differences between cholesterol and testosterone.

LET'S THINK ABOUT THIS

Homeothermic animals mostly contain saturated fatty acids. When living, their bodies are warmer than room temperature and the triglycerides stay in liquid form. Triglycerides with mostly unsaturated fatty acids are found in plants and poikilothermic animals. Hence we have beef fat, but corn oil and cod liver oil.

FUNCTIONS OF BIOLOGICAL MOLECULES

CARBOHYDRATES HAVE A VARIETY OF FUNCTIONS

Green plants trap light energy and use carbon dioxide and water to build the carbohydrate **glucose**. Glucose can be used directly or converted into larger carbohydrates. The functions of carbohydrates are shown in the table.

Carbohydrate	Function
glucose	broken down during respiration to provide chemical energy
starch	energy storage carbohydrate in plants
glycogen	energy storage carbohydrate in animals
cellulose	structural carbohydrate in plant cell walls

TYPES OF LIPIDS AND THEIR FUNCTIONS

The lipids are classified as biological molecules that are insoluble in water but which are soluble in organic solvents (such as ethanol). This definition, based on simple properties, brings together two completely different types of compounds: the triglycerides and the phospholipids which have a related structure and the steroids which are completely different. Types and functions of lipids are shown in the table.

Lipid type	Functions
triglycerides	long-term storage of energy (plants and animals)
	thermal insulation in homeotherms
phospholipids	structural, as part of the plasma membrane
steroids	hormonal, such as testosterone and oestrogen
	structural, as part of the plasma membrane

FUNCTIONS OF PROTEINS

Each gene carries the information required to code for one polypeptide. The polypeptide is a chain of amino acids which will become part of a protein. Many different proteins are made to keep a cell functioning. The table below describes some of the major functions of proteins.

Function	Examples of proteins
catalytic	enzymes, such as DNA polymerase (page 21 and 32) and kinase (page 28)
structural	tubulin in the cytoskeleton (page 25)
	collagen in bone and cartilage
messenger	hormones, such as ADH and insulin (page 30–31)
carriers	membrane proteins, such as aquaporin (page 23)

PROTEINS

STRUCTURE OF AMINO ACIDS

All amino acids have a central carbon which has **four groups** attached to it: a hydrogen; an NH_2 amine group at one end; a COOH carboxylic acid group at the other end; and a variable component, called an R group.

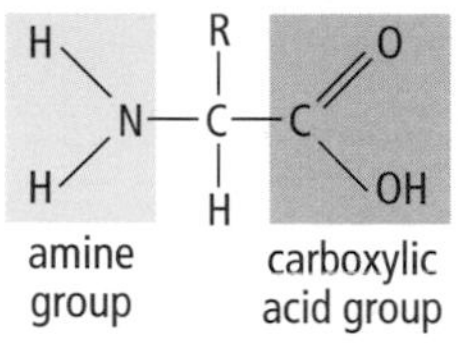

amine group

carboxylic acid group

The structure of amino acids

There are four classes of amino acids: **acidic**, **basic**, **polar** and **non-polar**. These classes are defined by the R group attached to the central carbon. The acidic, basic and polar R groups are **hydrophilic**; the non-polar R groups are **hydrophobic**. The properties of the R group are important in protein structure.

Amino acid class	acidic	basic	polar	non-polar
Interaction with water	hydrophilic	hydrophilic	hydrophilic	hydrophobic
Key component of R group	–COOH	$–NH_2$	–OH	hydrocarbon
Example	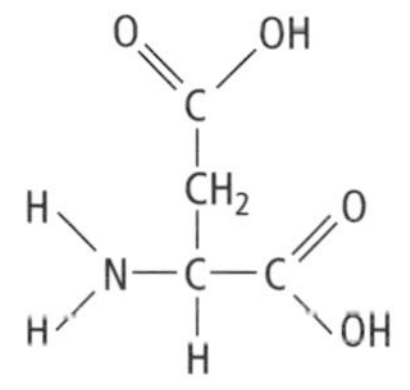 aspartic acid	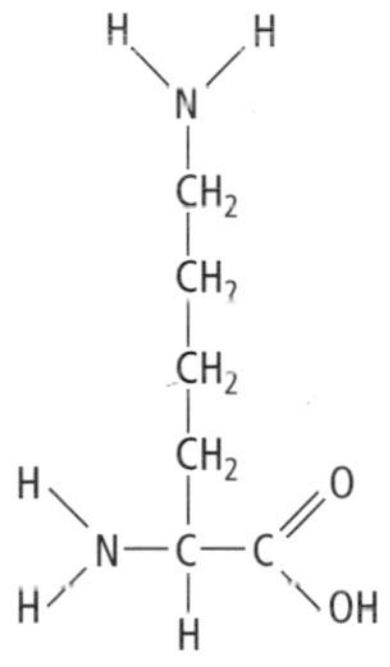 lysine	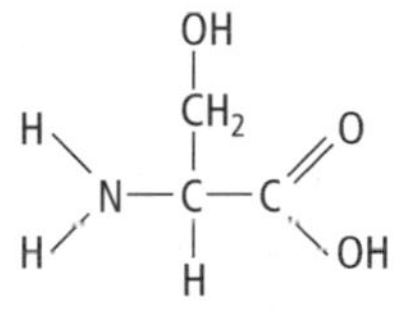 serine	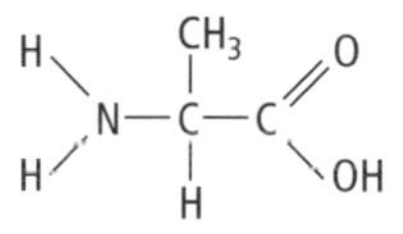 alanine

LINKING AMINO ACIDS

Amino acid monomers are linked together during translation of mRNA at the ribosome. An enzyme causes a **condensation reaction** between two adjacent amino acids. Water is removed by joining the OH group of the COOH of one amino acid to the hydrogen from the NH_2 of the other amino acid. The bond that links the amino acids is called a **peptide bond**.

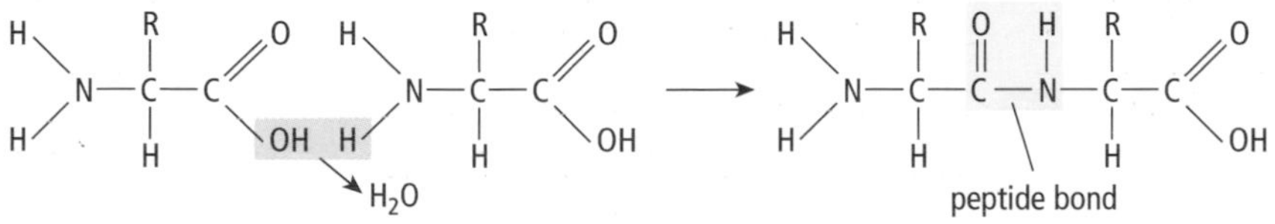

H_2O

peptide bond

A condensation reaction forming a peptide bond

DON'T FORGET

You do not have to learn the structure of any specific amino acids but you should know the general structure and the key features of the four classes of amino acids.

contd

PROTEIN STRUCTURE

Molecular biologists recognise four different levels in the structure of a protein.

> **DON'T FORGET**
>
> The primary structure determines the protein's secondary and tertiary structure.

Primary structure

This is the **sequence of amino acids** that make up a polypeptide chain. Each amino acid is linked to the next by a peptide bond to form a chain of amino acids. The chain has an **N-terminus** at one end and a **C-terminus** at other end.

N-terminus — H_2N–C–C–N–C–C–N–C–C–N–C–C–N–C–COOH — C-terminus

peptide bond

The primary structure of a small polypeptide. Can you identify the classes of the amino acids?

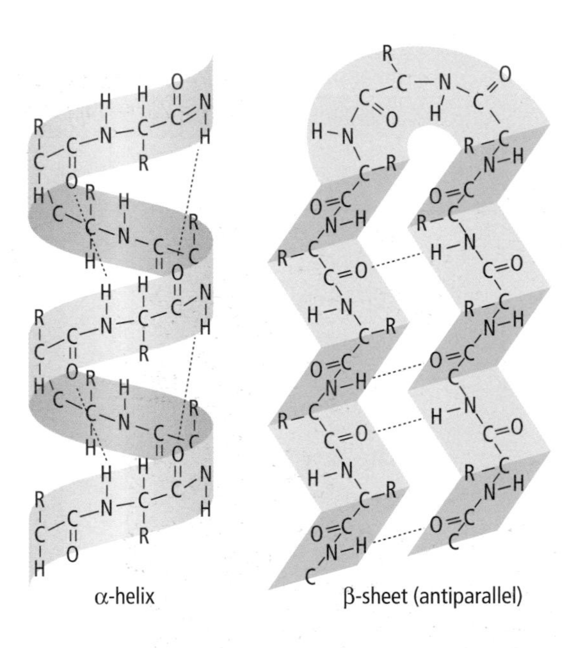

Secondary structure

Secondary structure is stabilised by **hydrogen bonds** between the atoms of different peptide bonds in the chain. The hydrogen of the N–H has a weak positive charge so it is electrically attracted to the weak negative charge on the oxygen of the C=O of another peptide bond.

There are two types of secondary structure.

- The **α-helix** is a spiral with the R groups sticking outwards.
- The **β-sheet** has parts of the polypeptide chain running alongside each other to form a corrugated sheet, with the R groups sitting above and below. The polypeptide chains are usually **antiparallel** (chains in **opposite directions** with respect to N-C polarity) but they can also be **parallel** (chains in the **same direction** with respect to N-C polarity).

Tertiary structure

This is the final 3-dimensional shape of the protein. It is stabilised by **interactions between R groups** of amino acids. The R groups were far apart in the primary structure, but the folding at the secondary level brings R groups close enough to interact.

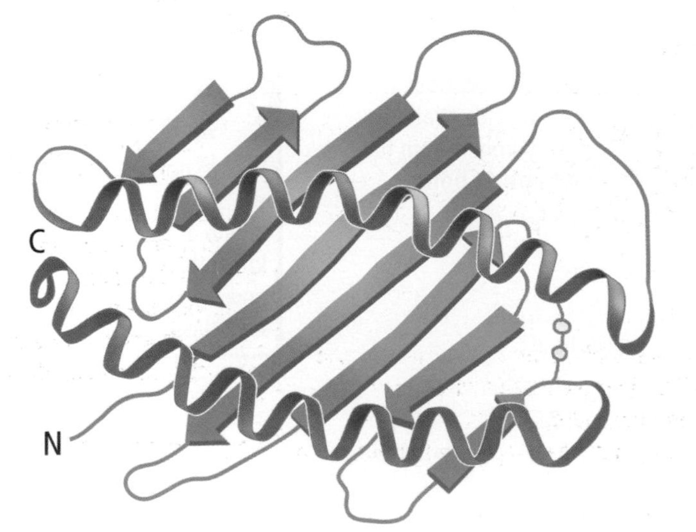

Tertiary structure of a protein. α-helix domains are shown in blue and antiparallel β-sheet domains in green. A disulphide bridge is also shown (yellow).

contd

PROTEIN STRUCTURE contd

Possible interactions between the R groups are shown in the table below.

Interactions between R groups	Description
hydrophobic interactions	Non-polar R groups are mostly arranged to the inside of the protein. The polar, acidic and basic R groups are mostly arranged on the outside.
Van der Waals interactions	Very weak attractions between the electron clouds of atoms.
hydrogen bonding	The weak negative charge of the oxygen of C=O is attracted to the weak positive charge on a hydrogen of an OH or NH_2 group.
ionic bonds	The COOH and NH_2 groups ionise to become COO^- and $NH3^+$. These groups are strongly charged and can attract each other.
disulphide bridges	Covalent bonds form due to reactions between the sulphur-containing R groups of cysteines.

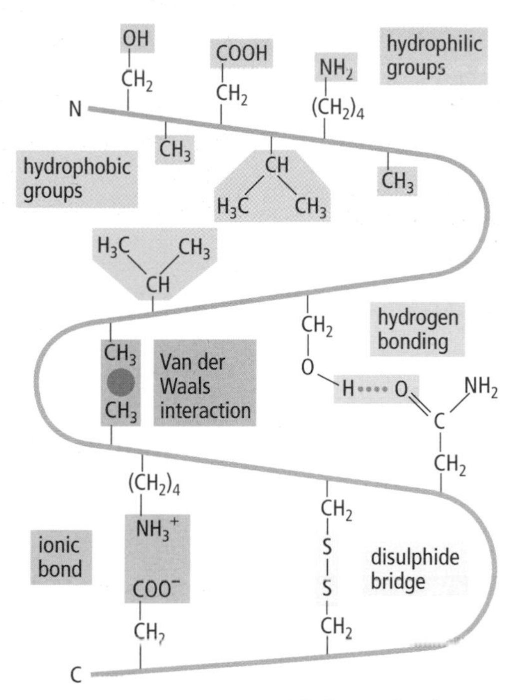

The interactions between R groups in tertiary structure

The final aspect of tertiary structure is the incorporation of **non-protein groups** into the polypeptide. These are called **prosthetic groups** and are very important for the function of the protein. For example, catalase enzyme has an iron atom at the heart of its active site.

Quaternary structure

Many proteins have more than one polypeptide **sub-unit**. These are linked by bonding between their R groups. For example, haemoglobin is made of four sub-units each with a prosthetic **haem group**.

The haem group of each sub-unit is shown in green.

The four sub-units of haemoglobin

Hair styling depends on changing the secondary and tertiary structure of the keratin protein. For a simple explanation of how this works, go to www.thenakedscientists.com/forum/index.php?topic=5911

DON'T FORGET

Secondary structure is due to hydrogen bonds between the atoms of the peptide bonds. Tertiary structure is due to interactions between the R groups.

LET'S THINK ABOUT THIS

Most solids become liquid when warmed. So why does the egg-white protein albumin go solid when it is heated? The energy from the heat shakes the albumin protein chain until the weak bonds of the secondary and tertiary structure are broken and the hydrophobic R groups are exposed. The protein chains are now repelled by water so they start to clump together forming a solid mass.

NUCLEIC ACIDS

THE INFORMATION MOLECULES

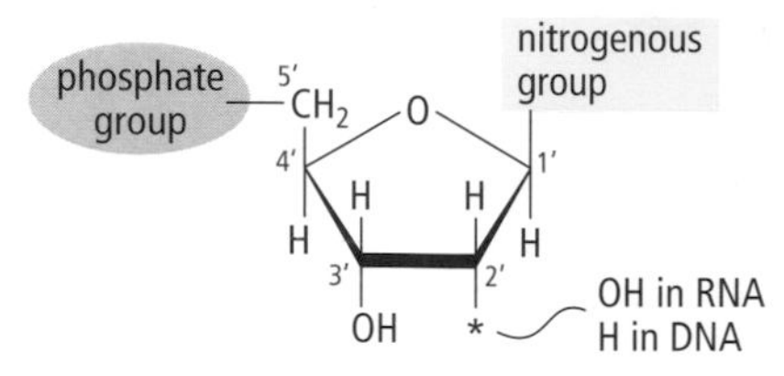

Simple structure of nucleotides

All genetic information is stored as a four-letter code in nucleic acids. All living organisms use DNA to store their information, while different viruses use DNA or RNA. The genetic information in DNA has to be copied to RNA before it can be read to form polypeptides and proteins. This has led to the theory that the genetic information of life started as an RNA system and evolution later 'backed-up' the information into DNA molecules.

Nucleotides

Nucleic acids are made from long chains of sub-units called **nucleotides**. These consist of a phosphate, a sugar and a nitrogenous base. The phosphate is attached to the sugar by a phosphoester bond. In RNA the sugar is **ribose** while in DNA the sugar is **deoxyribose**. The only difference between these two sugars is the H or OH attached to Carbon 2 (2′).

DON'T FORGET

Trying to remember that the **pur**ines are **a**denine and **g**uanine is **pure ag**ony.

DON'T FORGET

You only have to remember the number of rings, not the full structure!

The bases come in two different forms.

1. The **purines** are adenine and guanine and they have a **double-ring** structure.
2. The **pyrimidines** are cytosine, uracil and thymine and they have a **single-ring** structure.

DNA has the four bases adenine, cytosine, guanine and thymine. RNA has adenine, cytosine and guanine, but it has uracil instead of thymine.

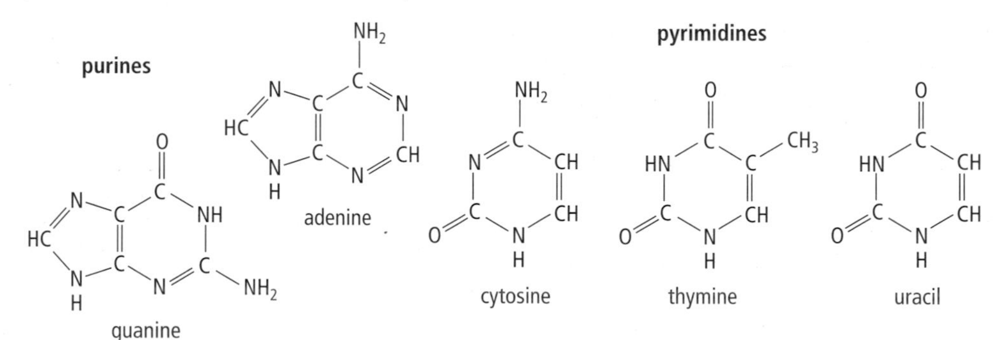

Purines and pyrimidines

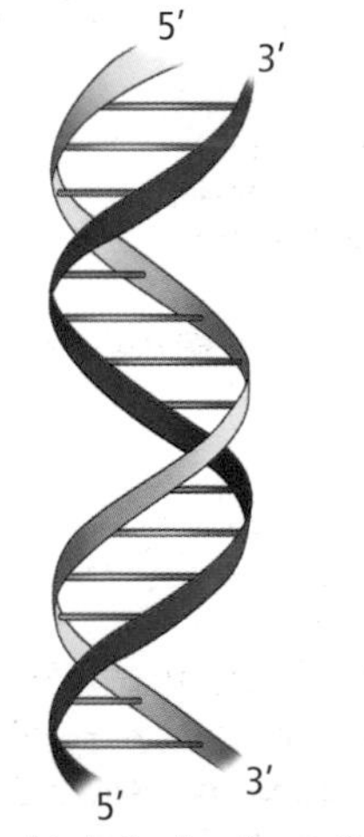

The double helix showing anti-parallel strands. Follow two strands from 5′ to 3′.

The DNA double helix

DNA is made up of two strands of nucleotides which are twisted together to make a twin spiral, or **double helix**. Each of these strands has a backbone made of deoxyribose sugars linked to phosphates by **phosphodiester** bonds. The phosphodiester bond links the phosphate of Carbon 5 (5′) to Carbon 3 (3′) of the next nucleotide. This means that strands of nucleotides have a **polarity** with 5′ at one end and 3′ at other.

The 5′ to 3′ polarity is important as the two strands of the double helix run in opposite directions – they are **antiparallel**. The two strands are linked together by 'rungs' made of a pair of bases, one from each strand. The bases in each pair are linked by **hydrogen bonds**. The base pair adenine and thymine is linked by **two hydrogen bonds**. Cytosine is linked to guanine by **three hydrogen bonds**.

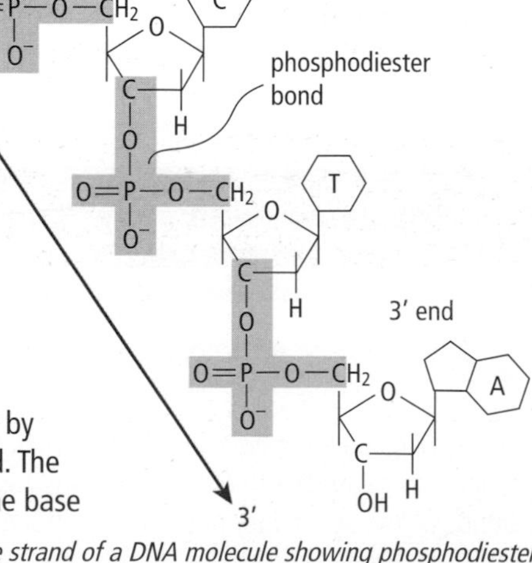

One strand of a DNA molecule showing phosphodiester bonds and the 5′ to 3′ polarity . Notice the negative charge on the oxygen attached to the phosphate.

DON'T FORGET

You need to remember the base pairing and the number of hydrogen bonds between the bases. **C-3-G** and **A-2-T** could help you do this.

contd

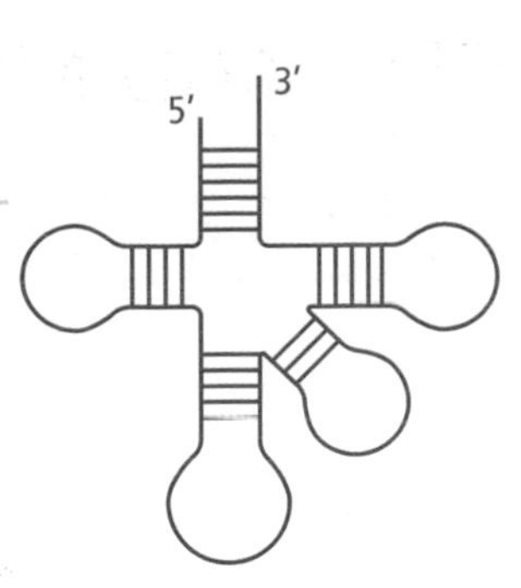

THE INFORMATION MOLECULES contd

RNA is a single-stranded molecule

In RNA, the nucleotide strand also has a 5′ at one end and 3′ at other. The strand may be looped back on itself to form base-paired regions, but it is still a single strand of nucleotides. For example, tRNA molecules have between 75 and 95 nucleotides and five regions of base pairing.

Diagram of a tRNA molecule showing the base paired regions within the single strand

COPYING STRANDS OF NUCLEIC ACID

New nucleic acid is synthesised for two distinctly different purposes:

- DNA replication during S phase of the cell cycle (see page 6)
- transcription of DNA to RNA during protein synthesis (gene expression).

Synthesis of a new strand of nucleic acid can only happen in the **5′ to 3′ direction** because the **polymerase enzymes** can only add new nucleotides on to the 3′ end of a nucleotide. The transcription of RNA only requires one strand to be made, so **RNA polymerase** can start at one point and build a 5′ to 3′ messenger RNA strand. But DNA replication requires both strands to be copied (one going 5′ to 3′, the other going 5′ to 3′), so something special has to happen.

How does DNA replication manage to copy both strands?

Enzymes open up a replication fork in the DNA double helix. The **leading strand** is then synthesised complete in the 5′ to 3′ direction as the fork opens up. Replication starts with the synthesis of a short **RNA primer**; this is replaced with DNA later. **DNA polymerase** adds a new DNA nucleotide to the 3′ end of the primer and so starts to build the new DNA chain. DNA polymerase matches free DNA nucleotides to their complementary bases on the template strand of DNA and then catalyses the formation of the **phosphodiester bond** between the end of the growing strand and the new nucleotide. This phosphodiester bond is formed when the OH of the 5′ phosphate and OH of the 3′ of deoxyribose are involved in a **condensation reaction**.

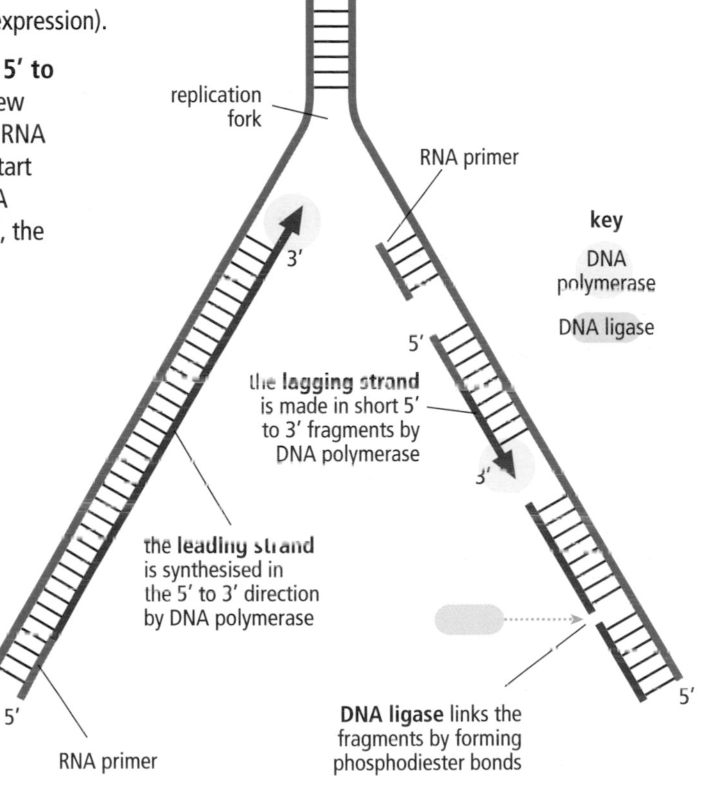

DNA replication

The other strand is known as the **lagging strand** and it is made in short 5′ to 3′ fragments (known as Okazaki fragments). As on the leading strand, RNA primers are made first and DNA polymerase adds nucleotides on to the 3′ end. After the RNA primer has been replaced with DNA, then **DNA ligase** links the 5′ to 3′ fragments of the lagging strand together by catalysing the formation of phosphodiester bonds between the nucleotides at the ends of the fragments.

DON'T FORGET

DNA polymerase and RNA polymerase can only build new strands of nucleotides in a 5′ to 3′ direction.

Search YouTube for *DNA structure garland* for a good 3.07 minute animated film showing many of the features of DNA structure, including the hydrogen bonding between the base pairs.

LET'S THINK ABOUT THIS

Many of the features of DNA have been harnessed in DNA technology.

1. In the correct conditions of pH, the OH unit in the phosphates dissociates to leave a negative charge. This means that the DNA has a negative charge and can be moved in an electrical field. This is the basis of gel electrophoresis.
2. Complementary base pairing allows gene probes to be made to bind to regions of DNA, so that they can be labelled and seen after electrophoresis.

MEMBRANES

MEMBRANES AND CELLS

Membranes – fluid bilayers of phospholipids studded with a mosaic of proteins – provide boundaries in cellular organisation. Selective permeability allows chemical gradients to be maintained across membranes, and the membrane surface is often the location of complex chemical reactions. To increase surface area, membranes are often highly folded, such as in the case of the cell membrane of certain **prokaryotes**. **Eukaryotic** cells (see pages 4–5) are distinguished by the presence of membranes within the cell itself. These membranes separate regions of differing biochemistry, and allow for the specialist functioning of the various components of the **endomembrane** system – the endoplasmic reticula, Golgi apparatus, vesicles, vacuoles and microbodies. Some organelles, such as the nucleus, chloroplast and mitochondria, are bounded by two membranes.

MEMBRANE STRUCTURE

The phospholipid bilayer

Phospholipids are composed of glycerol, two fatty acids and a phosphate; other groups, such as choline, can be attached to the phosphate 'head' (see page 15). The phosphate head is charged and therefore **hydrophilic**; the lipid 'tails' are non-polar and therefore **hydrophobic**. In aqueous environments, hydrophobic interactions cause phospholipids to form a **bilayer**, with the hydrophobic tails in the middle of the bilayer away from water and hydrophilic heads in contact with aqueous solutions. The hydrophobic centre of the phospholipid bilayer forms a barrier to the passage of polar molecules and ions.

Cholesterol controls membrane fluidity

The phospholipids in a bilayer are fluid and constantly change position. The fluidity of membranes is modified through the addition of cholesterol. Cholesterol has a hydrophilic hydroxyl 'head' (OH) and a rigid hydrophobic 'tail' (the four-ring steroid structure and non-polar side chain). The structure of cholesterol (see page 15) **reduces membrane fluidity** while also preventing lipid crystallisation at low temperatures.

The protein mosaic

Proteins are an important functional component of membranes. They can be peripheral or integral.

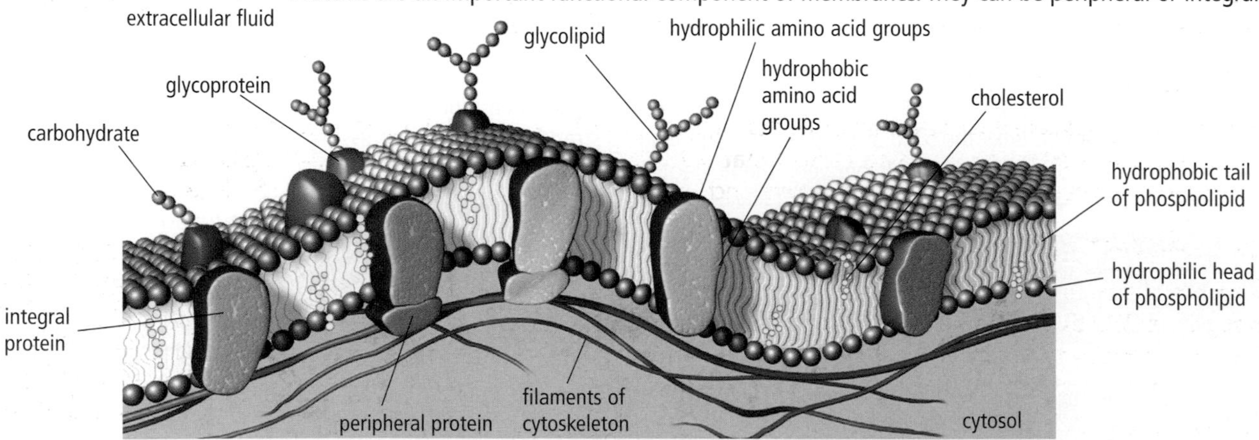

The fluid mosaic model of membrane structure

- **Peripheral** (or **extrinsic**) proteins are those that are easily removed from membranes in the laboratory using ionic washes. Peripheral proteins are only held in place at the surface of the membrane by charged or polar amino acid R groups or by a small number of hydrophobic interactions.
- **Integral** (or **intrinsic**) proteins are those that, in the laboratory, cannot be washed from the membrane. The integral proteins are held firmly in place within the membrane by strong hydrophobic interactions with the lipid tails. Integral proteins are either **transmembrane** (spanning the membrane and held in place by hydrophobic α-helices), or they are embedded in one side of the bilayer only.

MEMBRANE STRUCTURE contd

Glycoproteins and glycolipids

Some proteins and phospholipids have carbohydrate chains added to them so they become glycoproteins and glycolipids. The carbohydrate portions of these molecules are on the outside of the membrane.

FUNCTIONS OF MEMBRANE PROTEINS

Passive transport proteins

Passive transport proteins are transmembrane proteins that transport molecules across membranes **down a concentration gradient**. **Channel proteins** provide a pore that can facilitate or speed up diffusion. Each channel protein is specific for one particular ion or molecule; for example, aquaporin facilitates the diffusion of water across membranes. Transport can also be achieved by **carrier proteins** which bind to a specific molecule to allow its passage. Some of these transport proteins are 'gated' – the binding of one molecule is required to allow the passage of another.

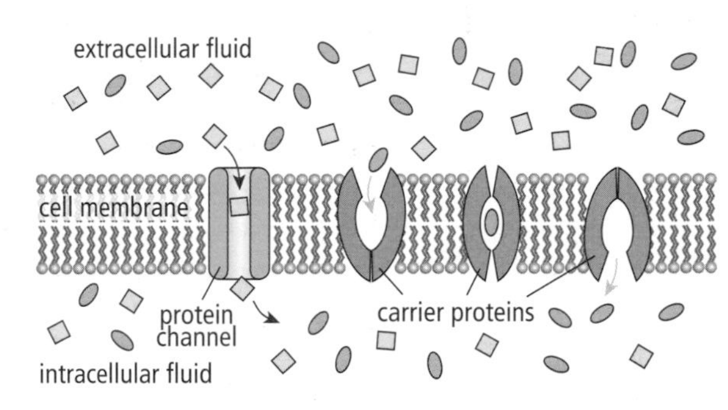

Passive channel proteins and carrier proteins moving molecules from high concentration to low concentration

Active transport proteins

Active transport proteins pump ions and molecules **against the concentration gradient**. The **hydrolysis of ATP** usually provides the energy source for the phosphorylation and conformational change of protein pumps. The sodium–potassium pump is a good example (see page 24).

extracellular fluid
sodium (Na^+)
potassium (K^+)
cell membrane
ATP
ADP
P_i
intracellular fluid
Na^+ K^+
– concentration +
+ concentration –

Active transport proteins using energy from ATP to pump molecules from low concentration to high concentration, as in this example of the sodium–potassium pump

Enzymes

Membrane-bound enzymes allow the location of catalysis to be carefully controlled within a cell.

Receptor proteins

Hydrophilic signalling molecules (see page 31) are unable to cross the hydrophobic region of the membrane. Specific receptor proteins are found in the plasma membrane of the relevant target cells. When signalled, these transmembrane receptor proteins stimulate a response within the cell, such as the phosphorylation of a key enzyme.

Attachment proteins

Proteins provide **cytoskeleton attachment points** within the membrane for the structural support of the cell. Other proteins **attach to the extra-cellular matrix** to hold the cell in place. In multicellular organisms, proteins form **intercellular junctions** that provide anchorage to other cells and hold cells together in tissues.

Cell–cell recognition

In multicellular organisms, cell–cell recognition is achieved by membrane glycoproteins. For example, the ABO blood grouping system in humans is based on the different carbohydrate chains presented by the glycoproteins on the membrane surfaces of red blood cells.

LET'S THINK ABOUT THIS

The double layers of membrane seen in mitochondria and chloroplasts in eukaryotic cells provide a strand of evidence that these organelles have arisen through endosymbiosis. The phospholipids of the internal membrane are like those of prokaryotes, while the outer membrane has phospholipids more like those of the eukaryotic cell. It is now accepted that the ancestors of mitochondria and chloroplasts were once free-living prokaryotes that became permanently engulfed by a larger cell. The nucleus also has a double-membrane structure and its origin is currently under debate.

THE SODIUM–POTASSIUM PUMP

Jens Skou won a Nobel prize for the discovery of $Na^+K^+ATPase$. Read a short autobiography at http://nobelprize.org/nobel_prizes/chemistry/laureates/1997/skou-autobio.html

See an animation at http://www.vivo.colostate.edu/hbooks/molecules/sodium_pump.html and find out how the sodium–potassium pump aids the absorption of glucose in your gut at academic.brooklyn.cuny.edu/biology/bio4fv/page/sympo.htm

DON'T FORGET

All animals rely on sodium–potassium pump pumping **2-K-in** (toucan).

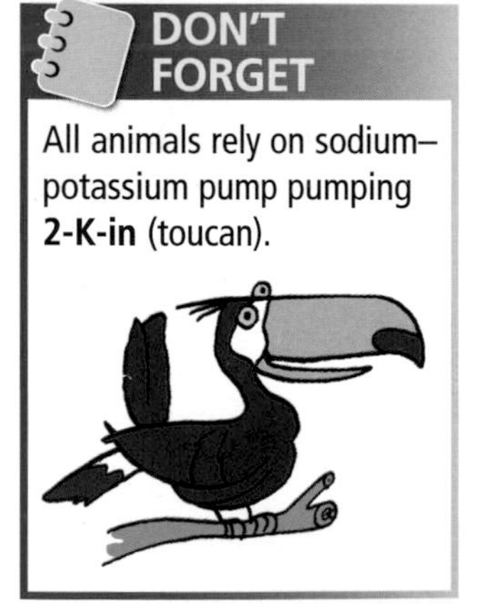

A VITAL TRANSMEMBRANE PROTEIN

The sodium–potassium pump (**$Na^+K^+ATPase$**) is a vital **transmembrane** ATPase commonly found in animal cells. The sodium–potassium pump moves sodium ions (**Na^+**) out of cells and potassium ions (**K^+**) into cells against steep concentration gradients.

The importance of the sodium–potassium pump

The sodium–potassium pump generates **concentration gradients** between cells and their environment. This is important in the gut, for example, where the sodium gradient aids glucose absorption. Note also that the sodium–potassium pump moves three positively charged ions out of the cell for every two that it pumps in. The **potential difference** that is created is essential for the generation of the resting potential in nerve cells. Some chemicals, such as digoxin from the foxglove, are inhibitors of the sodium–potassium pump. These chemicals act as potent toxins but can also have medicinal uses at very low concentrations.

How the sodium–potassium pump works

The sodium–potassium pump protein has **two stable conformational states**: one has a high **affinity for intracellular Na^+** and the other has a high **affinity for extracellular K^+**.

Stage 1: In one conformational state, the sodium–potassium pump has a high affinity for Na^+ ions. It exposes three Na^+ binding sites to the cytosol. Three Na^+ move in and bind to these sites.

Stage 2: When the three Na^+ are attached, the protein is able to hydrolyse an ATP molecule to ADP and phosphate. (This is why the protein is classified as an ATPase.) The phosphate is not liberated into the cytosol at this stage, but is bonded to part of the protein. This **phosphorylation** causes a **conformational change** to the protein.

Stage 3: The second conformation has a lower affinity for Na^+ ions but it can only release the Na^+ ions to the extracellular fluid, thus pumping the **three Na^+ out of the cell**.

Stage 4: The second conformation has a higher affinity for K^+ ions so K^+ from the extracellular fluid attach to the two K^+ binding sites. This triggers the release of the phosphate group from the protein.

Stage 5: The dephosphorylation restores the protein to its **original conformation**.

Stage 6: The original conformation has a low affinity for K^+ so it releases the **two K^+ into the cell**. The protein's affinity for Na^+ is high again and so the cycle repeats.

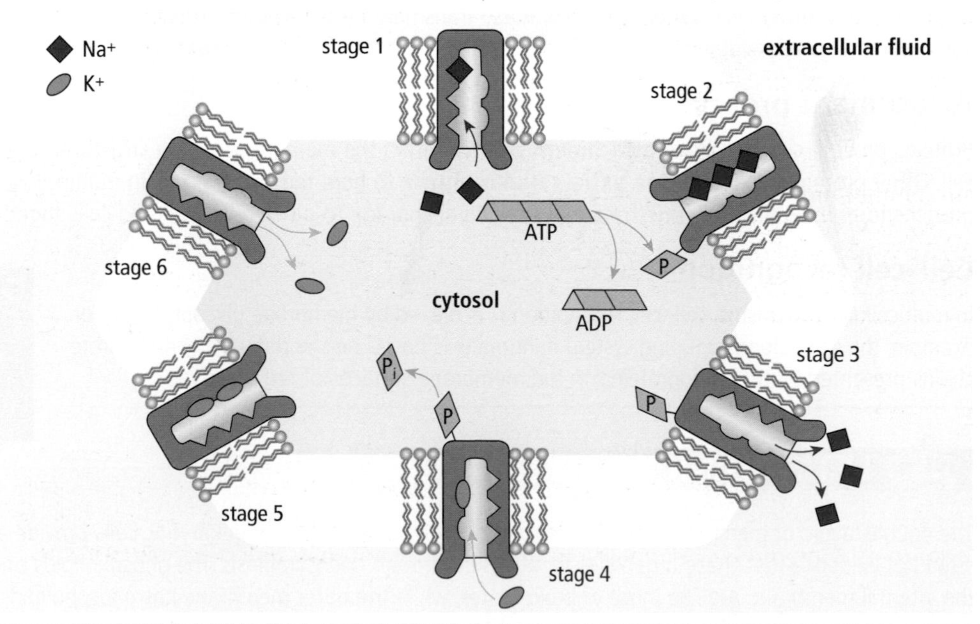

Six stages in the action of the sodium–potassium pump

THE CYTOSKELETON

THE CELL'S INTERNAL FRAMEWORK

Functions of the cytoskeleton

The cytoskeleton is a **network of protein fibres** that extends throughout the cytoplasm in all eukaryotic cells. The cytoskeleton is attached to membrane proteins (see page 23) and gives **mechanical support** to the cell, acting as scaffolding to maintain the shape of the cell. Organelles are attached to the cytoskeleton and it is involved in the **movement** of cellular components, such as vesicles or chloroplasts (movement of the latter can be seen under the light microscope). The cytoskeleton is also responsible for the movement of whole cells – pseudopodia, flagella and cilia all rely on cytoskeletal activity.

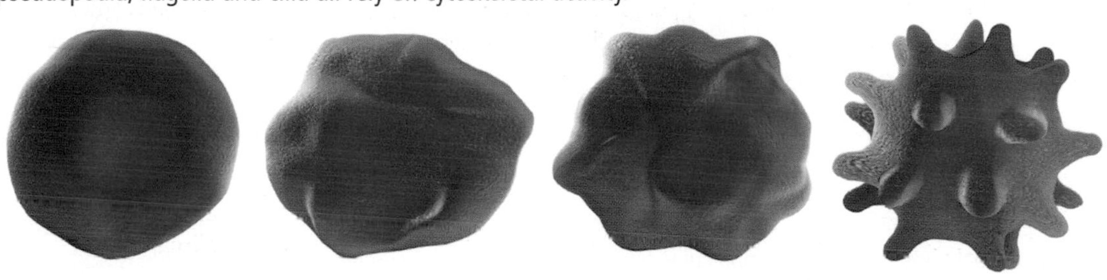

Progressive stages of crenation in a human red blood cell exposed to a hypertonic solution. As crenation develops, the internal scaffold of the cytoskeleton becomes obvious.

Microtubules

Microtubules are one component of the cytoskeleton. They are **polymers** of a dimer made from **α-tubulin** and **β-tubulin**; these are soluble globular proteins. The dimers are arranged to form microtubules which have a diameter of 25 nm. The length of microtubules is under the control of the cell through the addition (assembly) or removal (disassembly) of tubulin at the ends. Within a cell, microtubules originate from the **microtubule organising centre** (MTOC) which is located near to the nucleus and contains the **centrosome**. In animal cells, the centrioles are the site of microtubule synthesis within the centrosome.

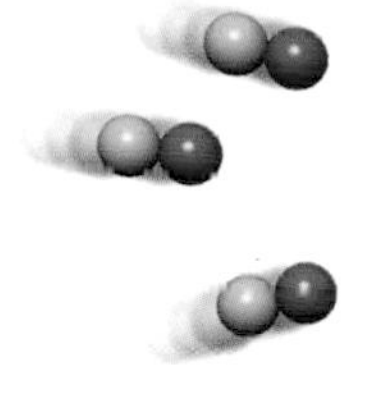

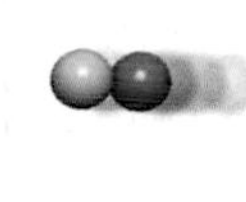

A microtubule is a polymer formed from tubulin dimers

Role of microtubules during cell division

Spindle fibres are made of microtubules – the cytoplasmic microtubules are disassembled to form the microtubules of the mitotic and meiotic spindles. Microtubules attach to proteins at the centromeres of chromatids. Microtubule severing **separates the chromatids**.

Read about research into how loss of a dynamic cytoskeleton may be implicated in skin wrinkling (www.newscientist.com/article/dn7173-wrinkles-could-be-less-than-skin-deep.html) and Alzheimer's disease (www.ncbi.nlm.nih.gov/pubmed/2874414).

LET'S THINK ABOUT THIS

Some drugs target microtubule action. Colchicine (from the autumn crocus) binds to soluble tubulin and prevents microtubule reassembly. It is used to induce polyploidy. Paclitaxel (from the yew tree) binds to tubulin in its polymer form and prevents the disassembly of the microtubules. It is used in chemotherapy as it triggers cell death in cancerous cells with abnormally dynamic microtubule function.

ENZYME ACTION AND INHIBITION

ENZYMES ARE BIOLOGICAL CATALYSTS

Activation energy

An enzyme binds to a substrate and stresses some of its chemical bonds, or binds to two substrates and forces them close together. These effects **lower the activation energy** needed to make a reaction happen. This means that the reaction is **catalysed** – it is much more likely to happen, so it speeds up.

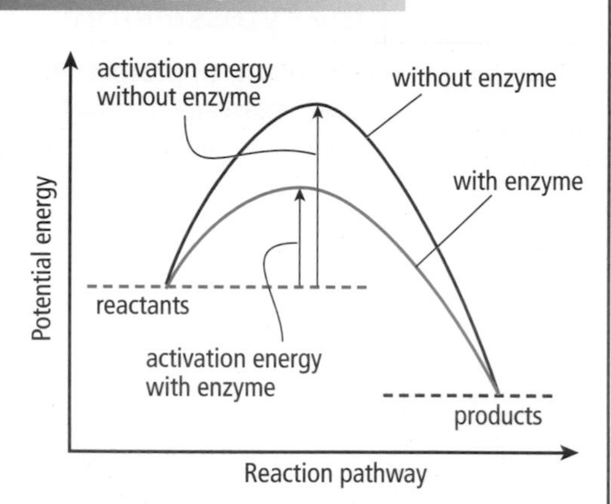

Metabolic enzymes

Enzymes regulate all the metabolic reactions in a cell. Some enzymes are involved in **synthesis** or **anabolic** reactions – these join molecules together, usually by a condensation reaction where a water molecule is removed. Other enzymes perform **degradation** or **catabolic** reactions which break molecules apart, often by hydrolysis where a water molecule is added.

DON'T FORGET

Metabolism is the sum total of all the anabolic and catabolic reactions.

HOW ENZYMES WORK

Enzymes have specific shapes

Most enzymes are made of protein. The amino acid sequence determines the shape of the protein molecule (see pages 18–19) and, as a result, also affects:

- the shape of the **active site** where the substrate binds
- which amino acids are present in the active site.

The picture of the primary structure of lysozyme shows that the amino acids of the active site are found far apart in the chain; the other amino acids provide the exact folding needed to hold the active site in position. Because of their highly defined structure, enzymes are **specific** to one substrate, or to a group of very similar substrates. For example, the enzyme **glucokinase** is specific to glucose, while **hexokinase** works on glucose and some other 6-carbon sugars.

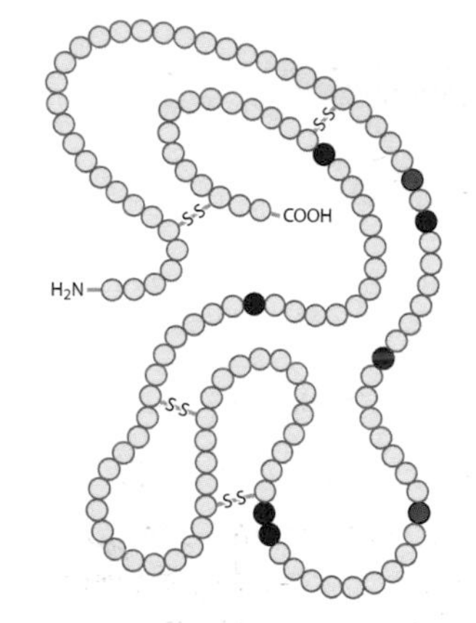

Primary structure of lysozyme. When folded into the tertiary structure (see page 19), the red amino acids bind to the substrate; black amino acids catalyse the reaction.

Induced fit

The lock-and-key hypothesis of enzyme specificity was put forward in 1894! The problem was that it didn't explain how the enzyme catalysed the reaction. In 1959, the **induced fit hypothesis** of enzyme action was developed. In this model, the shape of the active site still complements the shape of the substrate, but the active site also has amino acids with an **affinity** for areas on the substrate molecule. This means that the active site and the substrate bind by forming hydrogen bonds and ionic bonds.

The arrival of the substrate brings about a change in shape – a **conformational change** – in the enzyme due to the bonding between the active site and the substrate, which pulls the enzyme towards the substrate. The enzyme is showing an **induced fit** to the substrate. This makes the reaction more likely as the substrate is under tension – the activation energy of the catalysed reaction is lowered by stressing the bonds in the substrate.

DON'T FORGET

An active site has areas which match with the charge or polarity on the substrate.

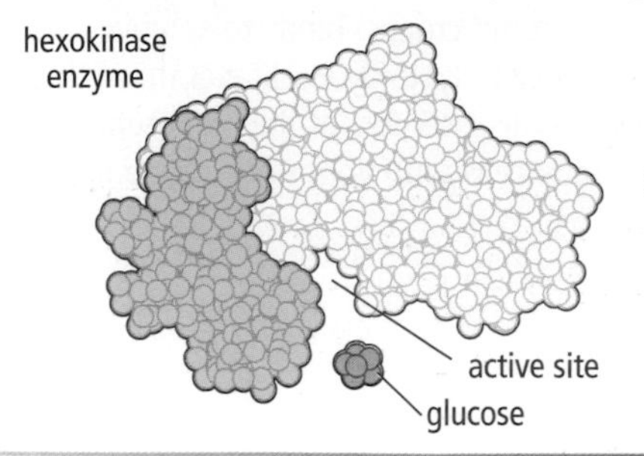

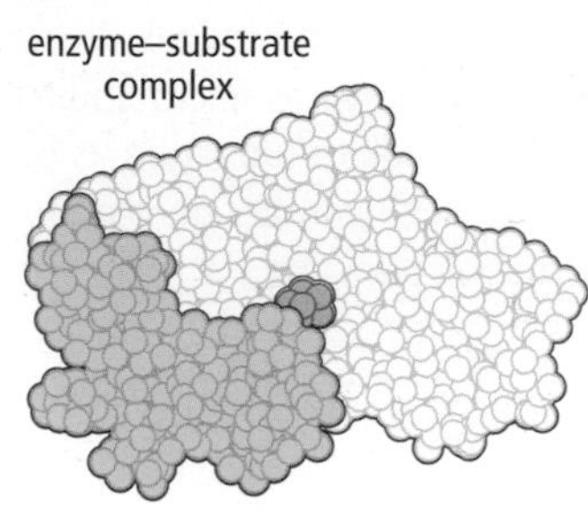

Hexokinase showing induced fit to glucose

ENZYME TYPES

There are four enzyme types that you are expected to know for Advanced Higher Biology.

Enzyme type	Action of enzyme	Examples
proteases	hydrolysis of peptide bonds to break down proteins	pepsin and trypsin from the digestive system bromelain from pineapples
nucleases	hydrolysis of phosphodiester bonds to break down nucleic acids	EcoRI used in genetic engineering to cut DNA
ATPases	hydrolysis of the phosphoester bond in ATP to form ADP and phosphate	the sodium–potassium pump in the membrane of cells (see page 24)
kinases	condensation reaction to add a phosphate group to another molecule	a kinase adds a phosphate to inactivate glycogen synthase (see page 28)

ENZYME ACTIVITY

The activity of an enzyme is found by using a constant concentration of enzyme and then measuring the initial rate of reaction at different substrate concentrations. As substrate concentration is increased, the initial rate of reaction increases until a maximum rate is reached (V_{max}).

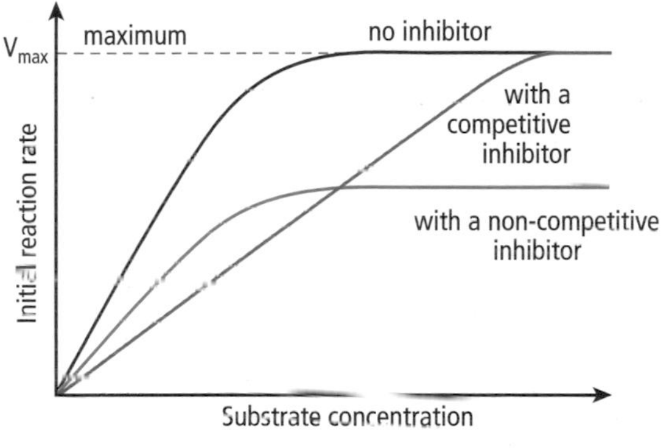

Inhibition of enzyme action

Inhibitors can reduce reaction rates in one of two ways: they can resemble the substrate, or they can alter the enzyme's shape so that it becomes inactive.

A **competitive inhibitor** is similar to the substrate in size, shape and charge pattern. If a fixed concentration of a competitive inhibitor is used when measuring enzyme activity, the increasing substrate concentration will eventually dilute the competitive inhibitor so much that all enzyme molecules bind to the genuine substrate and V_{max} is reached, albeit at a higher substrate concentration.

A **non-competitive inhibitor** binds to another part of the enzyme, away from the active site. Heavy metals (such as mercury and lead, see page 70) bind to the –SH groups of cysteines in the protein, and so change the enzyme shape. As shown in the diagram below, the active site no longer fits the substrate. If a fixed concentration of a non-competitive inhibitor is used when measuring enzyme activity, a proportion of the enzyme molecules are inactive, so V_{max} is reduced.

DON'T FORGET

Competitive inhibitors work because they are very similar to the substrate. These inhibitors are also likely to be specific to one enzyme.

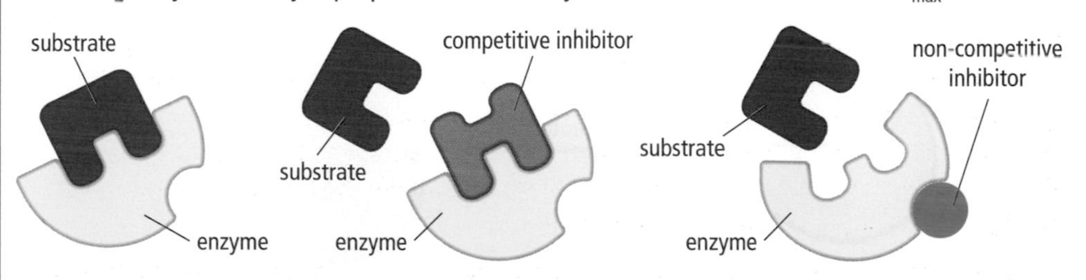

Enzyme interactions with substrates and inhibitors

Sarin gas was used in a terrorist attack on the Tokyo subway in 1995. Go to www.3dchem.com/molecules.asp?ID=25 to find out more about this poison.

LET'S THINK ABOUT THIS

Nitrogenase enzyme is the only enzyme that can fix atmospheric nitrogen for use in organisms (see page 52). The enzyme uses nitrogen gas (N_2) and can be competitively inhibited by the similar molecule, oxygen (O_2). Root nodules have leghaemoglobin to absorb oxygen and so prevent the inhibition.

ENZYMES IN METABOLISM

METABOLIC PATHWAYS NEED TO BE CONTROLLED

Metabolic pathways make essential products in the organism. But what if the cells have enough of a product already? What if more product is needed? There are mechanisms that control the activity of enzymes and so control the metabolic pathways.

COVALENT MODIFICATION

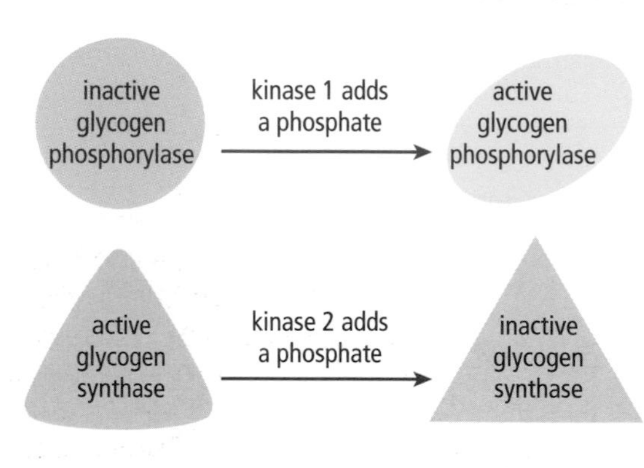

Phosphorylating cycle

Some enzymes can be activated or inactivated by having molecular units added or removed. In these examples, covalent bonds are made or broken, so this is called **covalent modification**. This changes the enzyme's shape so it is either made active or inactive. Either way, the concentration of active enzyme is changed, so the reaction rate is changed.

Kinase enzymes can **phosphorylate** other enzymes. The covalent addition of a phosphate by a condensation reaction **activates or inactivates** the enzyme. Similarly, phosphatase enzymes can **dephosphorylate** (remove a phosphate by hydrolysis) other enzymes causing them to be activated or deactivated. A good example of covalent modification is the control of glycogen metabolism.

When blood glucose concentration falls, two liver enzymes are phosphorylated by different kinases. One kinase enzyme adds a phosphate to **glycogen phosphorylase**; this activates to breakdown glycogen to release more glucose. At the same time, another kinase phosphorylates **glycogen synthase** so it is inactivated and stops synthesising glycogen. Obviously, when blood glucose concentration rises again, both these enzymes can be dephosphorylated to increase glycogen synthesis – another example of negative feedback in operation.

Another example of covalent modification is when covalent bonds in the polypeptide chain are cut, so changing the shape of the enzyme. Pancreas cells make an **inactive form** of trypsin called **trypsinogen**. This is secreted into the small intestine where it is cut by a **protease enzyme**. This converts the inactive trypsinogen into the **active form of enzyme, trypsin**. As the concentration of the active form increases, the reaction rate increases.

The addition or removal of a phosphate can activate or deactivate an enzyme, depending on how the shape of the enzyme is changed.

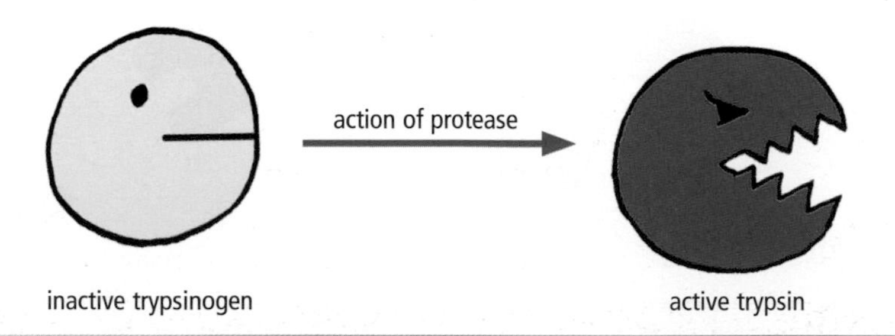

MODULATORS

The rate of formation of products by a metabolic pathway can be regulated by **changing the affinity** of one enzyme in the pathway for its substrate. This enzyme catalyses the **rate-limiting step** for the whole pathway and the affinity is changed by a **modulator molecule** that binds to the enzyme.

contd

MODULATORS contd

When a modulator binds, it **changes the shape of the enzyme**; for this reason the enzyme is called an allosteric enzyme ('other solid'). The modulator binds – through hydrogen bonds and ionic bonds – to the enzyme at a **second binding site** away from the active site. This is known as the **allosteric site.** The binding of the modulator causes a change in the enzyme's shape which, in turn, affects the shape of the active site and hence its affinity for the substrate. The rate of reaction will be changed as the effectiveness of the enzyme has been changed.

If the product of a metabolic pathway is in short supply, then a **positive modulator** (**activator**) can bind to the allosteric site and so increase the affinity of the enzyme for its substrate. This enzyme becomes more effective and makes more of the substrate for the next enzyme in the pathway, thereby increasing the rate of formation of all subsequent products in the pathway.

When there is enough of the final product in the pathway, then a **negative modulator** (**inhibitor**) can bind to a different allosteric site and so reduce the affinity of the enzyme for its substrate. The enzyme becomes less effective, makes less substrate for the next enzyme in the pathway and the whole pathway slows down.

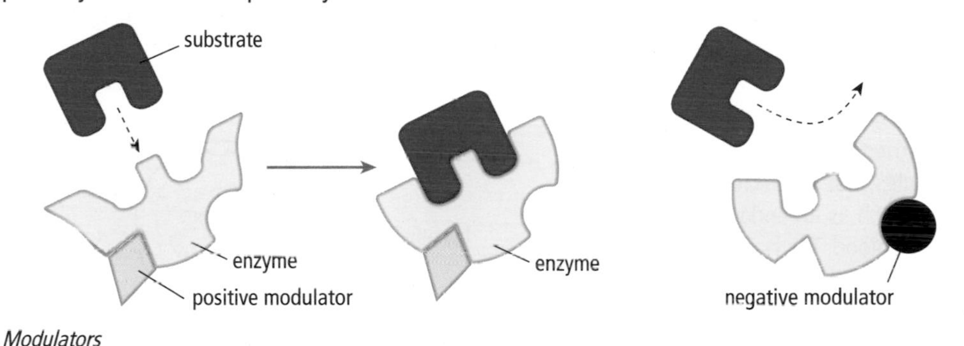

Modulators

DON'T FORGET

Modulators do not bind permanently. When their concentration in the cell falls, they are released from the allosteric site – so the enzyme's activity changes back to what it was before.

End-product inhibition

Many pathways are regulated by **end-product inhibition.** When the product of the final reaction accumulates, it slows down the first enzyme of the pathway and so slows down its own synthesis. In some pathways the final product works as a **competitive inhibitor** of the first enzyme, while in other pathways the final product is a **negative modulator** of the enzyme.

DON'T FORGET

This system is a form of **negative feedback**, with the pathway being regulated by the concentration of its end product.

In the example below, the end-product isoleucine is the negative modulator that binds to the allosteric enzyme at the start of the pathway. This means that, if the cell has excess isoleucine, it does not waste energy synthesising any more. And when the isoleucine concentration drops, the molecule is released from the allosteric site so the enzyme becomes more effective again and the pathway starts to produce more isoleucine again.

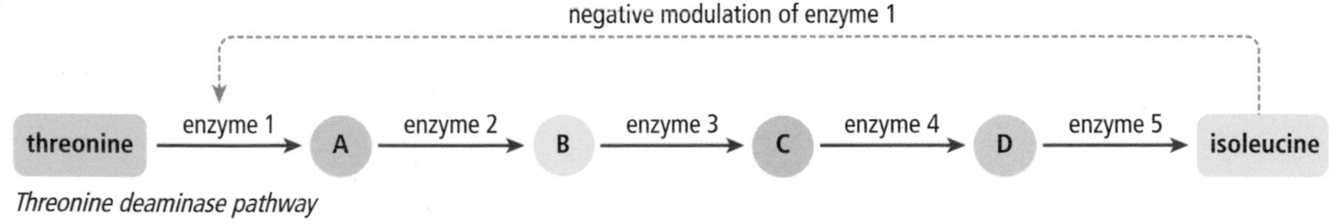

Threonine deaminase pathway

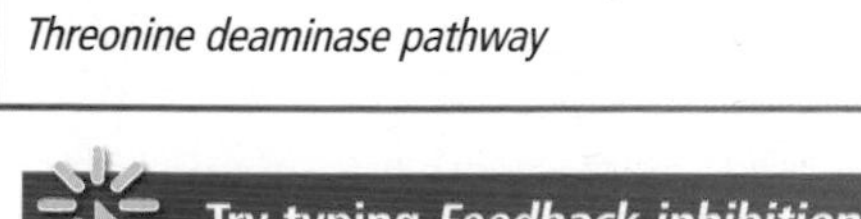

Try typing *Feedback inhibition of biological pathways* into YouTube for a 1.15 minute clip about this topic.

LET'S THINK ABOUT THIS

Why do pancreas cells make inactive trypsinogen? If they made active trypsin, they would be at risk of digesting themselves; having an inactive form means that the pancreas cell is safe. The enzyme only becomes active in the small intestine which has a mucus coating to prevent the proteases acting on the gut lining.

EXTRACELLULAR SIGNALLING

THE IMPORTANCE OF EXTRACELLULAR SIGNALLING

In a multicellular organism, it is essential that cells communicate with one another. Without this communication, the integration and coordination of cellular activities would be impossible.

The two principal forms of communication in multicellular organisms are **hormonal** and **nervous**. Both require the release of extracellular signalling molecules – that is, one cell releases a chemical signal molecule that another cell detects and responds to. These cells may be close neighbours or may be in quite different locations within the organism. Either way, the signal has to leave one cell and travel to another – the target cell. As these signals originate from outside the target cell, they are described as **extracellular signalling molecules**.

The same signal molecule can stimulate different target cells in different ways, and the same target cell may have various receptors, each specific to a different signalling molecule, therefore allowing the cell to respond to many different instructions.

Hormones

Hormones are extracellular signalling molecules that are secreted by one tissue (such as an endocrine organ) into the blood. The hormone circulates in the bloodstream until it reaches its target receptor or is broken down. The signalling mechanism of a hormone falls into one of two broad categories:

- **hydrophilic** hormones are the **peptide hormones** such as insulin, ADH and growth hormone
- **hydrophobic** hormones are the **steroid hormones** such as testosterone.

DON'T FORGET

A hydrophobic signalling molecule is able to pass through membranes. A hydrophilic signalling molecule cannot.

Neurotransmitters

The signalling molecules released during nervous communication are known as **neurotransmitters**. Neurotransmitters are **hydrophilic** molecules and they are released into the synaptic gap between a nerve cell and its neighbour. In comparison with hormonal communication, nervous communication is very specific due to this intimate association between the signalling cell and the target cell.

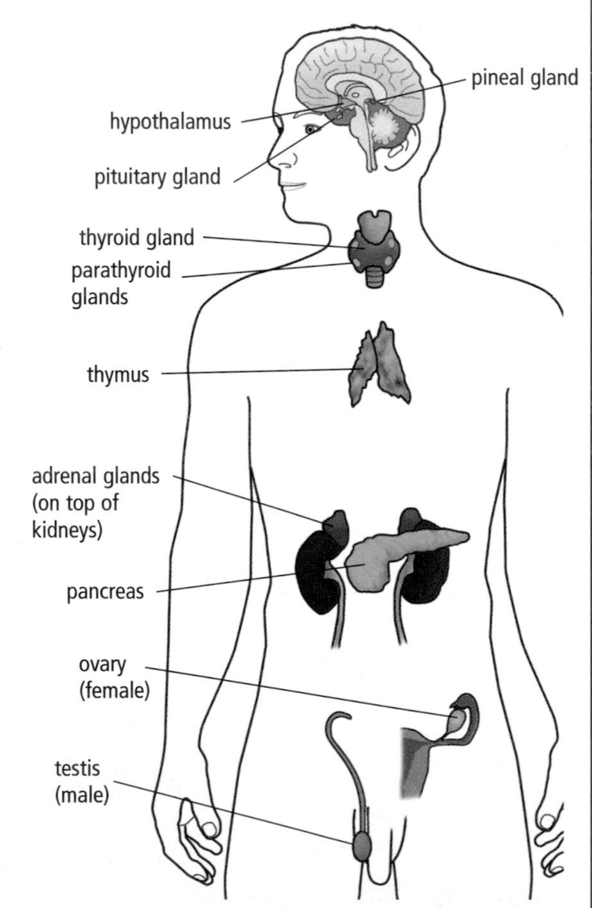

The main hormone-secreting endocrine organs in the human body

Location of receptors

An extracellular signalling molecule causes an effect when it **binds** to its specific target **receptor** molecule. Because **hydrophobic** signalling molecules can pass through the membrane, their receptor molecules are within the cytoplasm or the nucleus of the target cell. **Hydrophilic** signalling molecules cannot pass through the membrane so their receptors are integral proteins (see page 22).

HYDROPHOBIC EXTRACELLULAR SIGNALLING

Hydrophobic molecules are lipid soluble so they are able to move by diffusion across the hydrophobic part of the plasma membrane. The receptors for hydrophobic extracellular signalling molecules are, therefore, deep in the cytoplasm or nucleus of the target cell. The receptor for a steroid hormone is a **gene-regulatory protein** which changes conformation when the steroid binds to it. This conformation change activates the protein so that it can bind directly to the DNA and regulate the transcription of specific genes.

Steroid hormones

There are many examples of hydrophobic signalling molecules, such as **testosterone**, oestrogen and progesterone. Each one shares the common four-ringed structure of steroids but varies in its side chains (see page 15).

HYDROPHILIC EXTRACELLULAR SIGNALLING

Hydrophilic molecules are not lipid soluble and, therefore, cannot pass through the hydrophobic barrier of the cell membrane. Instead, they bind to a transmembrane protein receptor at the cell surface. Once the signal molecule binds to the protein at the surface of the cell, the conformation of the transmembrane protein alters in some way. This results in **signal transduction** into the cell, changing the behaviour of the cell. An ion channel at the cell surface (see page 23) may open or a secondary messenger within the cell cytosol may be released. Secondary messengers can alter cell behaviour by activation of proteins by kinase enzymes (see page 28) and phosphorylation, or through pathways involving calcium ion release from the endoplasmic reticulum.

Search on YouTube using the terms *apoptosis signal transduction* for an amazing animation of a transduced signal resulting in cell death

Peptide hormones

Peptide hormones are small hydrophilic proteins. Well-known examples include **insulin**, **glucagon**, **anti-diuretic hormone** (ADH) and **somatotrophin** (growth hormone). Each one requires a specific receptor protein at its target cell surface. Since only target cells have the appropriate receptors at their surface, the action of these hormones can be highly specific.

Neurotransmitters

Acetylcholine (ACh) and **noradrenalin** (NAd) are examples of neurotransmitters. Both are hydrophilic peptides. Acetylcholine is the transmitter at the neuromuscular junction connecting motor nerves to muscles. Noradrenaline has a role in the central nervous system and the sympathetic nervous system. Unusually, noradrenaline can also act as a stress hormone if released into the blood.

Noradrenaline can act as a neurotransmitter and also as a stress hormone, such as in the fight-or-flight response both in this long-tailed macaque and the human behind the camera!

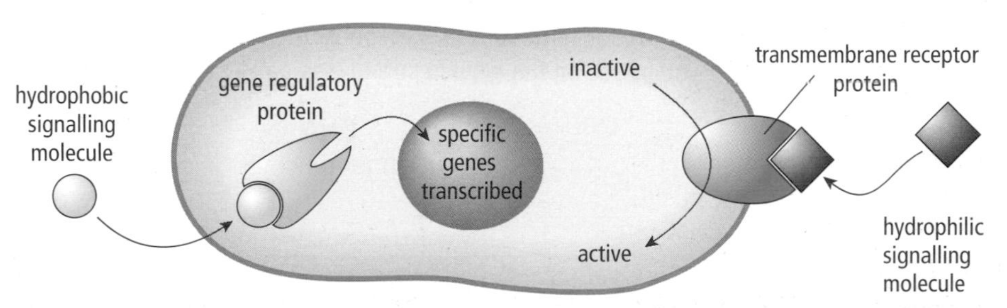

A summary of extracellular signalling pathways

Search for a 4:12min YouTube clip on *Insulin Production and Type 1 Diabetes* for a great animation of insulin receptors at work. For an introduction to the role of acetylcholine as a neurotransmitter, watch www.youtube.com/watch?v=hzXVe4RS8-A

LET'S THINK ABOUT THIS

Since the 1923 Nobel Prize for Medicine was awarded to Banting and Macleod, many Nobel prizes have been awarded for research into aspects of cell signalling. To get a flavour of some of the challenges, rivalries and successes involved in the discovery of insulin, watch the feature film *Glory Enough for All*.

DNA ENZYMES

The manipulation of DNA in the laboratory relies heavily on the catalytic properties of various enzymes. For laboratory use, the enzymes are manufactured using the biotechnological processes shown on pages 78–79.

POLYMERASES

Polymerases catalyse the **condensation** synthesis of DNA and RNA **nucleic acid** polymers. Hydrogen bonding between complementary pairs of monomer nucleotides and the base sequence of a template strand organises the nucleotides into the correct position and sequence. Polymerases catalyse the formation of a **phosphodiester** bonds between a nucleotide and the nucleic acid of the newly synthesised strand. This new strand can be synthesised by polymerase in a 5′ to 3′ direction only (see page 21).

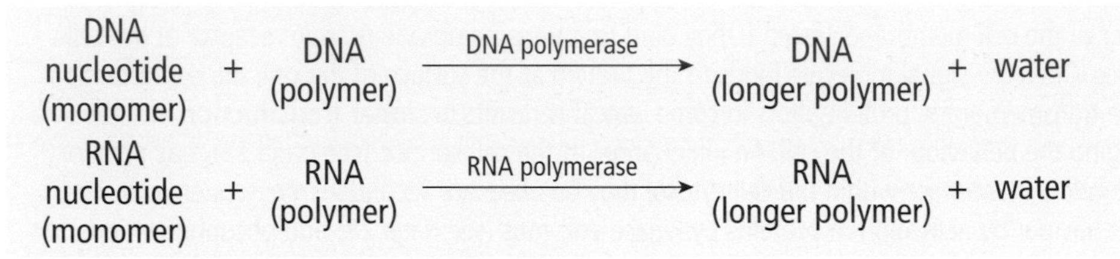

The action of DNA polymerase and RNA polymerase

DON'T FORGET

When discussing polymerase, remember to specify whether you are referring to DNA or RNA polymerase.

Since polymerases catalyse the addition of monomers (nucleotides) onto the end of a strand of polymer (the nucleic acid), a short sequence of polymer complementary to the start of the template strand is required to initiate polymerisation. These short single-stranded complementary sequences are known as **primers** (see pages 21 and 36).

In the cyclic reactions involved in **PCR** (the polymerase chain reaction; see page 36) and DNA sequencing (see pages 36–37), a thermostable DNA polymerase must be used. This DNA polymerase has been cloned from a hot-spring bacterium, *Thermus aquaticus*, and is stable to 95°C. It is usually referred to as Taq polymerase.

DNA LIGASE

DON'T FORGET

Polymerase and ligase are anabolic enzymes, whereas endonucleases are catabolic enzymes.

Ligase catalyses the **condensation** synthesis of **phosphodiester** bonds between two strands of nucleic acid polymer. In DNA replication, it joins the sugar–phosphate backbone between fragments of DNA as they are synthesised on the lagging strand (see page 21).

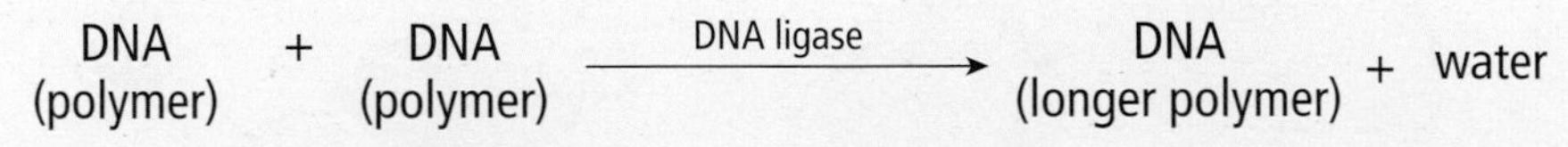

The action of DNA ligase

ENDONUCLEASES

In general, endonucleases catalyse the **hydrolytic** degradation of phosphodiester bonds. In digestion, they catalyse the complete breakdown of nucleic acid polymers into nucleotide monomers.

DNA (polymer) + water —endonuclease→ DNA nucleotides (monomers)

The action of endonuclease

contd

ENDONUCLEASES contd

Restriction endonucleases

Restriction endonucleases (often referred to as restriction enzymes) also break the phosphodiester bonds between adjacent nucleotides in a sequence but they are *restricted* to making this cut between a specific sequence of nucleotide bases.

DNA (polymer) + water —restriction endonuclease→ DNA (shorter polymer) + DNA (shorter polymer)

The action of a restriction endonuclease

Sticky ends and blunt ends

All restriction enzymes cut phosphodiester bonds at specific nucleotide base sequences. Some restriction enzymes cut straight across the DNA strand, breaking the two phosphodiester bonds opposite one another and producing DNA with what are known as **blunt ends**. However, most restriction enzymes make a staggered cut in DNA, hydrolysing a phosphodiester bond on each strand several base pairs apart. This staggered cut is known as a **sticky end**.

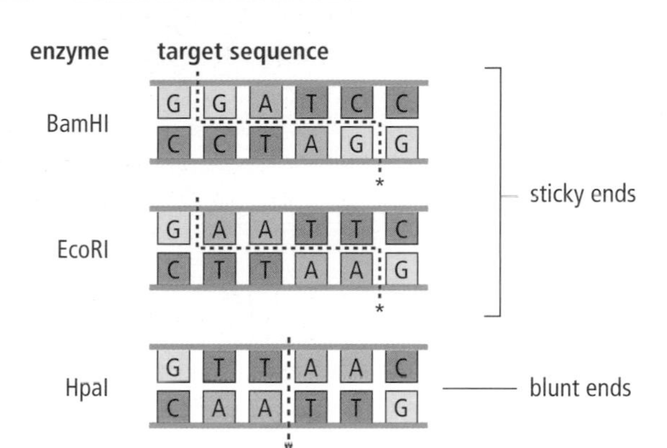

The action of restriction enzymes is restricted to specific nucleotide sequences.

An advantage of blunt-ended DNA fragments is that they can be joined to any other blunt-ended fragment. An advantage of sticky ends is that the short sections of single-stranded DNA at the ends only form hydrogen bonds with complementary single-stranded sequences. This means that DNA fragments with sticky ends will hybridise with other DNA fragments produced by the *same* restriction enzyme, but not with fragments produced by a different restriction enzyme.

PLASMIDS AND RECOMBINANT DNA

Plasmids are small loops of DNA found mainly in prokaryotic cells. Prokaryotic cells exchange plasmid DNA in their natural environment, making plasmids ideal both as **vectors** for the movement of genes between cells in the laboratory, and the manufacture of **recombinant DNA**. The term recombinant DNA refers to combinations of genes or DNA that would not occur naturally.

Plasmids have a variety of uses in the laboratory. They are used to transform prokaryotic cells to produce useful proteins, such as human insulin or growth hormone. The **Ti plasmid** is used as a vector to transform eukaryotic plant cells (see page 45). Plasmids of different sizes are also used to store libraries of DNA fragments within both prokaryotic and yeast cells (see page 39). Plasmids used in biotechnology are generally engineered to have an area with multiple restriction-enzyme cut sites. If a plasmid is cut open in this region with a particular restriction enzyme producing sticky ends, any other fragment of DNA cut open with the same restriction enzyme will have complementary sticky ends and can easily be sealed into the plasmid using ligase. With knowledge of the position of the various possible restriction-enzyme cut sites in a particular plasmid, the precise locations for insertion of fragments of DNA can be controlled. Recombinant plasmids tend to be engineered to include several genes other than the fragment of interest. These genes generally include an antibiotic resistance or fluorescent protein, so that cells which have taken up the engineered plasmid can be selected easily.

LET'S THINK ABOUT THIS

Restriction endonuclease enzymes evolved in prokaryotes as a defence against bacteriophage virus attack. Restriction enzymes cut DNA at specific sequences that are not found in the bacterial genome but that are found in bacteriophage DNA.

DNA GELS, BLOTS AND PROBES

GEL ELECTROPHORESIS

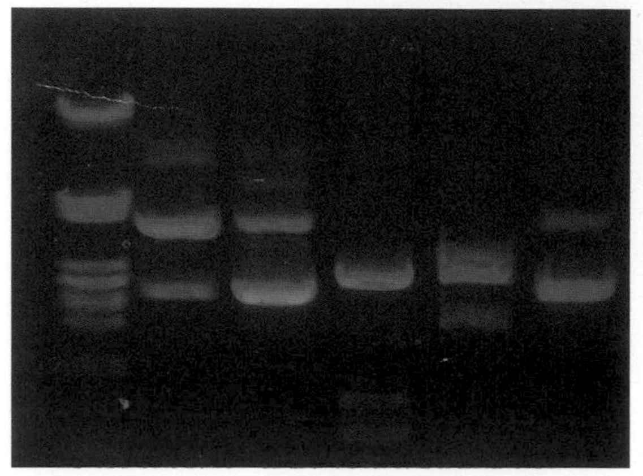

DNA is separated into bands of different sizes in the lanes of a gel electrophoresis. The left-hand lane contains the DNA ladder for reference.

When DNA is digested by a restriction enzyme, fragments of different lengths are produced. DNA fragments of different lengths can be **separated** using gel electrophoresis. The digested DNA sample is placed into a well in a porous polysaccharide gel immersed in a buffer solution. An electrical current runs between two electrodes placed in the buffer. DNA is negatively charged, so moves towards the positive electrode. As the charge is uniformly distributed along the sugar–phosphate backbone of the molecule (see page 20), the DNA fragments are separated along their lane according to their size, with shorter fragments moving faster and therefore further through the gel. One well in a gel would usually contain a DNA ladder – this lane then displays a selection of known lengths of DNA that can be used as a reference scale.

DENATURING AND ANNEALING DNA

DON'T FORGET

There are 2 hydrogen bonds between A and T but there are 3 between C and G.

The hydrogen bonds holding the complementary bases together in DNA can be broken by **heating** the DNA to approximately 95°C. This is known as **denaturing** or melting the DNA and results in single-stranded DNA.

When the temperature is lowered, complementary single-stranded sequences form hydrogen bonds to remake a double helix. This formation of a double-stranded molecule from single strands is known as **hybridising** or **annealing** DNA.

DNA sequences rich in C:G complementary pairs have more hydrogen bonds than sequences rich in A:T pairs. So, the melting temperature of DNA is higher in fragments with a higher proportion of C:G pairs.

DNA BLOTTING

Electrophoresis gels easily shrink and deform, and so are unsuitable materials for further processing or storage of DNA. To make an accurate record of the final positions of the DNA fragments after electrophoresis, DNA blotting procedures are used to draw the DNA fragments out of the gel and onto a nitrocellulose or nylon filter. Often, the DNA is denatured chemically with sodium hydroxide before this procedure to enhance the attachment of the DNA to the filter.

restriction fragment preparation

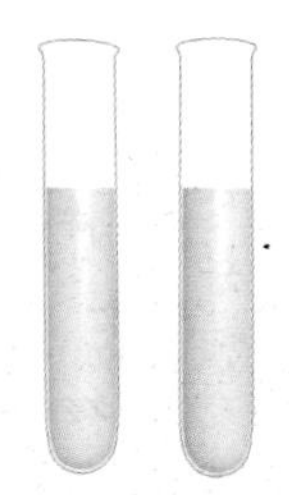

DNA samples are incubated with restriction enzymes to produce fragments

gel electrophoresis

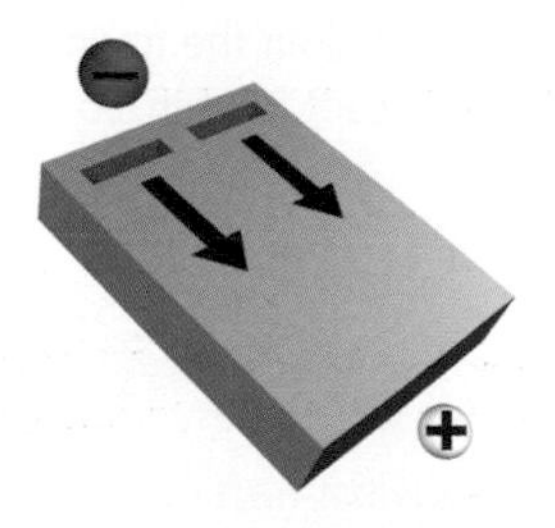

DNA fragments separated by current according to size

blotting

DNA denatured then drawn up through gel onto filter placed on top of gel

probing

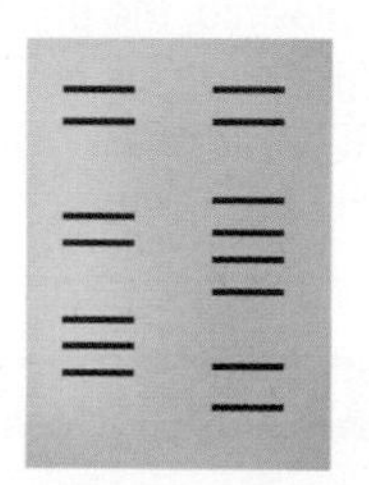

The probe hybridises with any complementary sequences. The sizes of the fragments can be compared.

The Southern blot procedure combines a restriction enzyme digest, gel electrophoresis, blotting and probing.

DNA PROBES

Probes are **short single-stranded sequences of DNA** that have been labelled either radioactively or with a fluorescent tag. Their sequence is designed to be complementary to a target DNA sequence.

Typically, the results of a blot would be incubated with a probe. The probe binds with any **complementary sequences** of DNA that are present in the sample. The blot is then washed to remove any unbound probe. The bands where the probe has bound are revealed using photographic film or light of a specific wavelength.

Single-locus probes

Single-locus probes are those that have a complementary sequence at only **one location** within the genome being probed. This means that they can only hybridise at this one location. A band on a single-locus probe gel will indicate the presence of the sequence in the fragment at that location in the gel. Single-locus probes can be used for paternity tests and to screen for mutations within a gene sequence (see page 41).

Multi-locus probes

Multi-locus probes are those that have a complementary sequence which occurs at **many locations** within the genome being probed. If a genome is digested using a restriction enzyme, a multi-locus probe will hybridise at many locations and produce a pattern of bands known as a DNA profile or fingerprint. Variation in the size of the fragments containing sequences complementary to the probe results in differences in the profiles produced by different individuals or organisms. The original forensic profiles were made using multi-locus probes.

DNA microarrays

DNA microarrays are a more recent development in DNA probe technology. In a DNA microarray, thousands of different probes are attached individually into a series of micro wells. Multiple copies of the DNA sample under test are amplified (often cDNA from the reverse transcription of RNA). The DNA is tagged using fluorescence and then flooded into the wells. After washing, the wells with hybridised DNA can be detected using fluorescence. DNA microarray technology allows a sample of DNA to be tested by many thousands of different probes at one time.

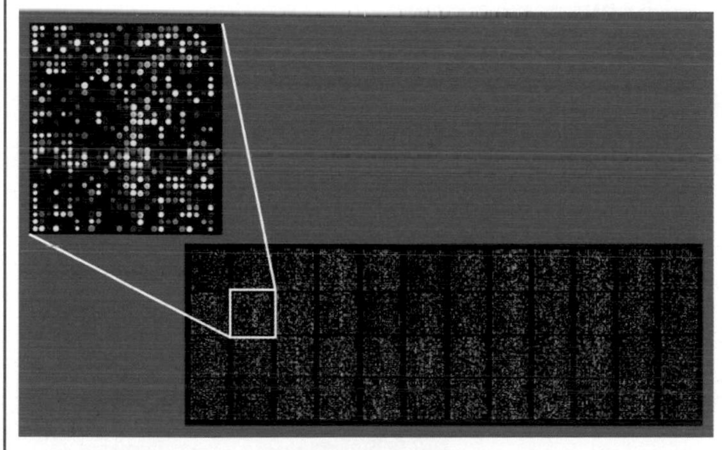

A microarray of around 40,000 different probes

LET'S THINK ABOUT THIS

Probes and primers are both short single-stranded complementary sequences of DNA. Probes are used so that *we* can identify the location of a specific sequence. Primers are used so that *DNA polymerase* can locate a start point for DNA replication.

PCR AND DNA SEQUENCING

THE POLYMERASE CHAIN REACTION (PCR)

The polymerase chain reaction (PCR) is used to **amplify** DNA fragments. Amplification refers to the generation of many identical copies of a particular DNA fragment. PCR involves a cycle of reactions at three different temperatures. By the end of each cycle, the number of copies of the template DNA fragment has doubled.

DON'T FORGET

Primers anneal because they form hydrogen bonds with their complementary base pairs.

Initial requirements

PCR requires template DNA, the four DNA nucleotides, thermostable Taq polymerase (see pag 32), buffer and an automated reaction vessel known as a thermal cycler.

A sample of DNA is needed to act as the template for amplification. A small sample of DNA isolated from a trace of blood, skin, hair, semen or cheek cells is sufficient. Two primers target the section of DNA to be replicated. Each one is carefully selected to be complementary to a short stretch of bases on one of the two DNA strands. The primers bind to the template and allow DNA replication to occur in a 5′ to 3′ direction (see page 21). As each cycle progresses, the DNA between the two primer binding sites will be replicated.

Thermal cycling

The thermal cycler is a machine that automatically controls the thirty to sixty cycles of the following three different temperatures. As each thermal cycle doubles the number of copies of the target DNA, the number of copies of DNA increases exponentially. After 30 cycles, it is possible to achieve amplification to a billion copies!

- **Stage 1 – Denaturation at 95°C**
  The mixture is heated to 95°C to denature the DNA by breaking the hydrogen bonds between the bases. Taq polymerase is thermostable and does not denature.
- **Stage 2 – Annealing of primers at 55°C**
  Cooling allows the primers to anneal (form hydrogen bonds) to their complementary base sequences on the single-stranded DNA.
- **Stage 3 – DNA extension at 72°C**
  The optimum temperature for the Taq DNA polymerase to extend the complementary strands in a 5′ to 3′ direction is 72°C.

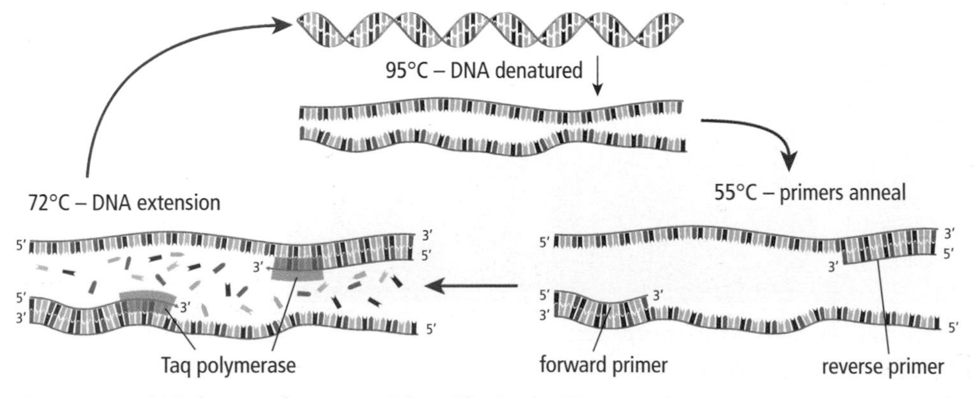

Primers target a DNA fragment for exponential amplification by PCR.

DNA SEQUENCING

DNA sequencing is the determination of the **order of the nucleotide bases** in a fragment of DNA. To determine the positions of each of the four bases in a fragment, a methodology that is very similar to PCR is used, with the exception that **chain-terminating** dideoxynucleotides are added to the mixture.

contd

DNA SEQUENCING contd

Dideoxynucleotides

Dideoxynucleotides differ from deoxynucleotides in that they lack a hydroxyl (OH) group on Carbon 3. The lack of this OH group prevents further phosphodiester bond formation, so no more extension of the DNA polymer occurs. As the dideoxynucleotide is the last nucleotide that can be added on this particular fragment, it is known as a chain terminator.

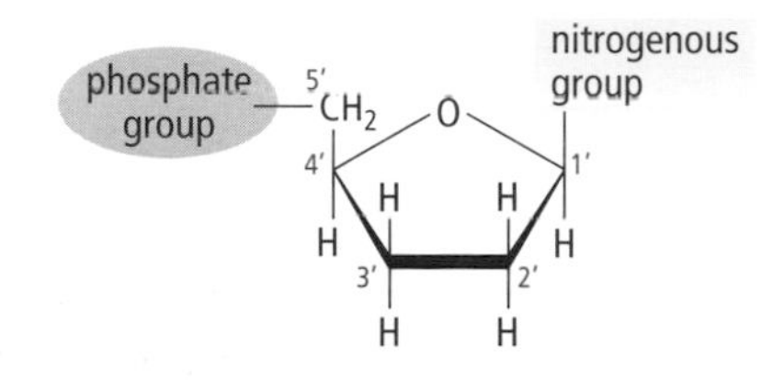

Structure of a chain-terminating dideoxynucleotide. Note the lack of a hydroxyl group at carbon 3.

The fragment of DNA to be sequenced is replicated with an appropriate ratio of free nucleotides to chain-terminating dideoxynucleotides. This ensures that fragments of every possible length are produced – from only one base added to the primer, to the full length of the fragment.

DON'T FORGET

Compare the structure of this dideoxynucleotide with the other nucleotides on page 20.

Two methods of sequencing

In the **Sanger sequencing method**, the primer is radioactively labelled. Four separate PCR-type incubations are carried out. In each one, a different chain-terminating dideoxynucleotide is included (A, T, C or G). The different chain terminators produce a different selection of fragment lengths in each of the four incubations. The fragments produced are run in four adjacent lanes (one for each base) in a gel electrophoresis. Once the result has been blotted and exposed to photographic film, the base sequence can be read by eye from the bottom up.

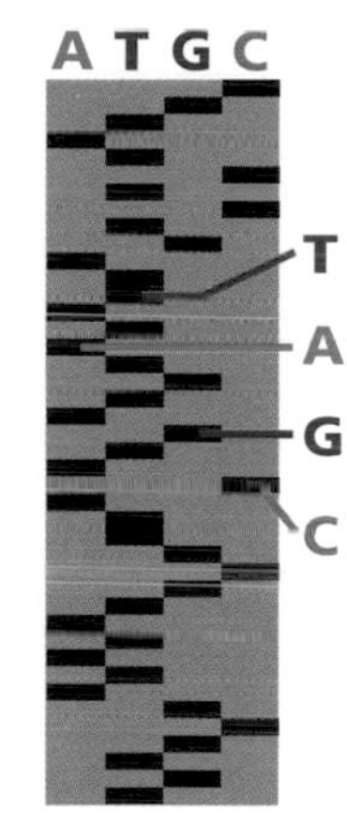

Reading the result of Sanger method sequencing

In the **dye-terminating sequencing method**, each of the four chain-terminating dideoxynucleotides is tagged with a different colour of fluorescent dye. This means that each unique length of fragment can only have one fluorescent tag – the one on its final chain-terminating dideoxynucleotide. The different fragment lengths are separated using gel electrophoresis and the fluorescent tag on the final base is detected by an optical reader.

The diagram below shows the stages in the dye-terminating sequencing method.

- **Stage 1** – DNA replication in the presence of the four dideoxynucleotides and four deoxynucleotides amplifies the DNA to make fragments of every possible length.
- **Stage 2** – Each fragment length is terminated by a fluorescent-tagged dideoxynucleotide.
- **Stage 3** – The fragments are separated according to size by gel electrophoresis.
- **Stage 4** – A detector identifies the colour of the fluorescence of each fragment, and thereby the sequence of the fragment.

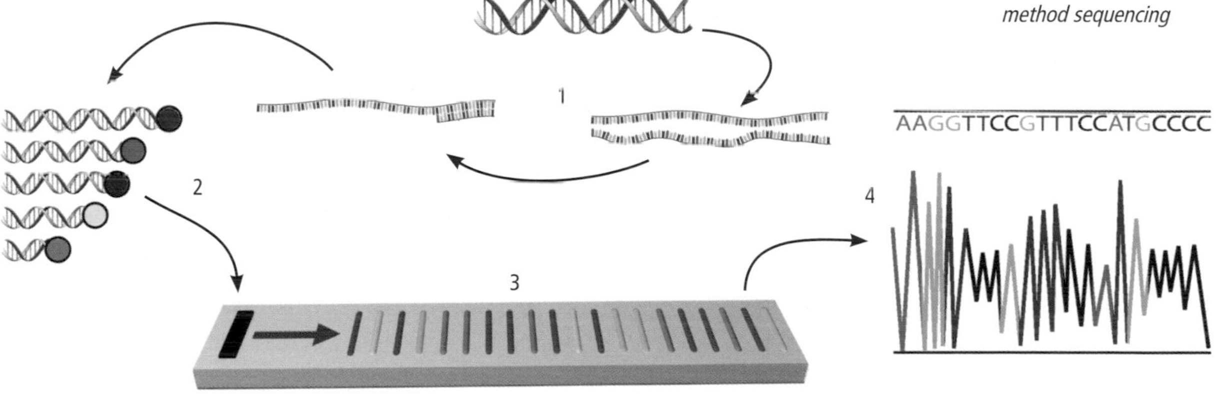

Dye-terminating DNA sequencing

Watch a 4:37 YouTube clip on PCR to see how DNA can be amplified millions of times

LET'S THINK ABOUT THIS

Advances in automation and electronic technology have greatly enhanced the speed of both PCR and DNA sequencing, reducing their costs enormously. For example, the costs of sequencing are estimated to be halving every two years or so. As both PCR and DNA sequencing become even cheaper, more and more applications will be developed.

GENOME MAPPING

The development of rapid DNA sequencing technology has allowed the entire genomes of many organisms to be sequenced and the positions of their genes to be mapped on their chromosomes.

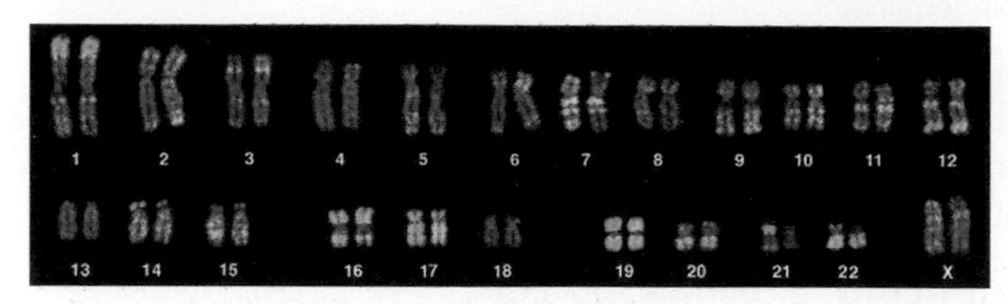

Human chromosomes

GENETIC LINKAGE MAPPING

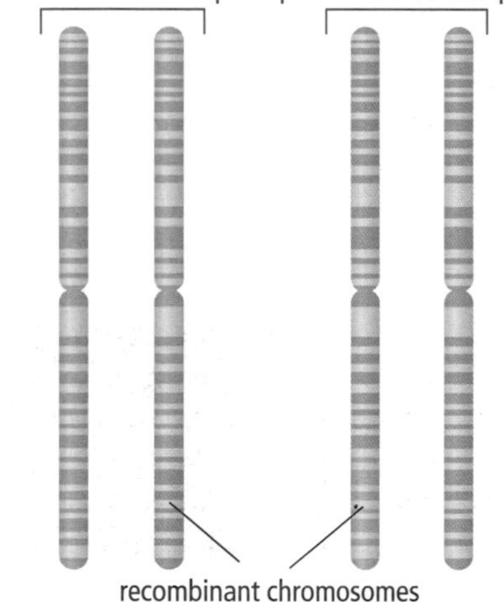

Crossing over of homologous chromosomes during meiosis causes recombination of alleles.

Traditional linkage mapping has been used since the 1920s to map chromosomes. Initially, the technique had to rely on the alleles of genes coding for visibly variable characteristics in an organism's phenotype. The **recombination** of alleles is a result of crossing over at meiosis, and pairs of genes that are further apart are more likely to form recombinations compared with pairs of genes that are closer together.

More recently, linkage mapping has been used to map the inheritance of many other types of **genetic marker**, such as probe-binding sites and restriction enzyme cut sites. In this way, the frequency of recombination between these known genetic markers and genes or other genetic markers can be used to build a genetic linkage map.

While linkage maps are very useful, they do have some drawbacks. They rely on pedigree-type information, or at least access to gametes. More importantly, the chances of crossing over are not uniform along the length of a chromosome. Crossing over hot-spots occur and, in humans, recombination is more frequent in females. For these reasons, genetic linkage maps do not measure the physical distance between the markers being mapped.

PHYSICAL MAPPING

Physical maps are designed to determine the **order of the genes** on each chromosome and the **distance between them** in terms of the number of DNA base pairs (bp). The ultimate physical map is the complete sequence of bases. Before the full sequence is attained, a physical map should ideally consist of a series of sequence tagged sites (STS). Each of these sites consists of a unique base sequence found nowhere else in the genome being mapped. In addition, the distance in terms of bp along the chromosome should be known for each STS. As an STS is easily probed for, it can act as a signpost for the addition of more detail on the physical map.

DON'T FORGET

A genetic linkage map is based on frequency of recombination between two markers. A physical map is based on the number of base pairs between two markers.

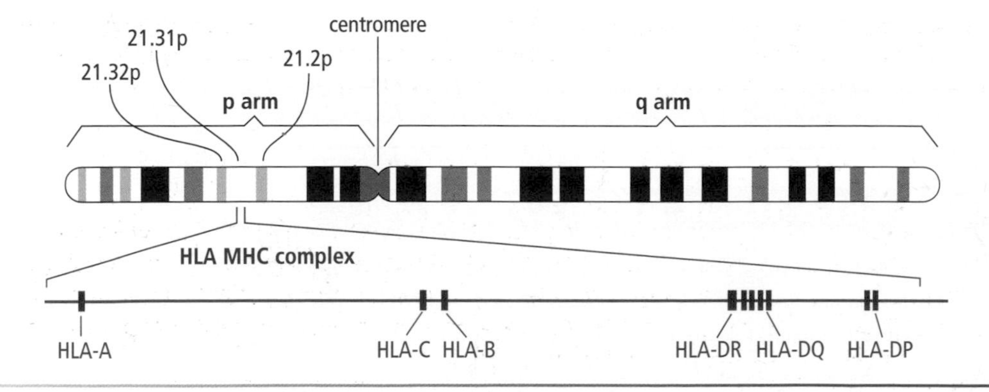

A selection of physical markers on part of the map of human chromosome number 6

HUMAN GENOME PROJECT

The Human Genome Project was an international scientific collaboration that began in 1990 and was completed by 2003. In that time, the 3 billion bases of the human genome were read on average nine times; as a result 99% of the human genome is now publicly available. The accuracy of the sequence is estimated to be 99.99%. (The remaining more-difficult parts of the sequence are the repetitive areas associated with centromeres and telomeres, and particular coding sections relating to the immune system.)

The project involved much development of techniques, as well as systems for sharing research information. Once everything was in place, progress was rapid, with 80% of the sequence being published in one single year.

Clone libraries

Initially, all the human chromosomes were broken by restriction enzymes into fragments about 150 000 bp in length, each with a unique STS. Each fragment was stored within an *E. coli* bacterium as an artificial chromosome or plasmid. Each stock of bacteria replicated its fragment as it reproduced – each stock was a clone library for one particular fragment and could be frozen until required for sequencing.

Chromosome walking

Before sequencing, each fragment had to be broken into smaller 2000 bp lengths. This was done several times, using separate restriction enzymes. The fragments from the different restriction enzyme digests were sequenced and then compared by computer. Since the restriction enzymes cut each batch at different points in the overall sequence, the regions of overlap could be determined.

Genomes of other species

Genomes of other species have been sequenced and many genes have been found to be the same as in humans. Some species can be used as models for research on the influence of genes on an organism. Yeast has been used in the study of the cell cycle. The roundworm *C. elegans* is used to study cell death in development. The mouse is used as a mammalian model for human genetic conditions.

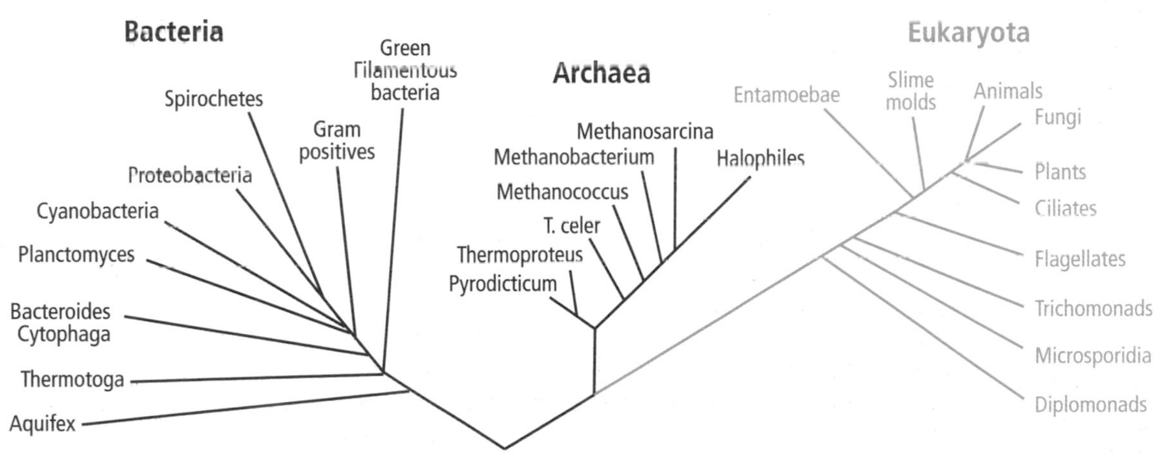

Genome comparison studies have shown life to be divided into three main domains, the Bacteria (such as E. coli*), the Archaea (such as* Thermus aquaticus*) and the Eukaryota (such as protoctistans, fungi, plants and animals).*

LET'S THINK ABOUT THIS

The comparison of the human genome with other species reveals remarkable similarities. Many gene sequences, such as the *Hox* genes that control developmental segmentation in animals as different as humans and *Drosophila*, have been conserved through long periods of evolutionary change. Through the comparison of whole-genome data, the evolutionary relationships of living organisms are being determined much more precisely.

DNA TECHNOLOGY AND THERAPEUTICS

GENETIC DISORDERS

A **genetic disorder** is a disease caused by an abnormality in an individual's DNA. The central dogma of biology emphasises the link between DNA and protein. In a genetic disorder, mutant alleles are responsible for the synthesis of abnormal proteins which then result in disease.

In the field of medicine, DNA technology can be used for screening, diagnosis and treatment. **Screening** involves the identification of individuals who are carriers of particular genotypes or markers that are associated with an increased risk of disease. **Diagnosis** involves procedures that identify the presence of disease. Trial **treatments** using DNA include gene therapy.

DON'T FORGET

A change to the primary structure of a protein is likely to change its function.

CYSTIC FIBROSIS (CF)

Cystic fibrosis is a genetic disease caused by a mutant form of the CFTR protein. The non-mutant form of this transmembrane protein contains 1480 amino acids. It controls the chloride-ion balance in the cells lining the lungs and digestive system by channeling chloride ions (Cl^-) out.

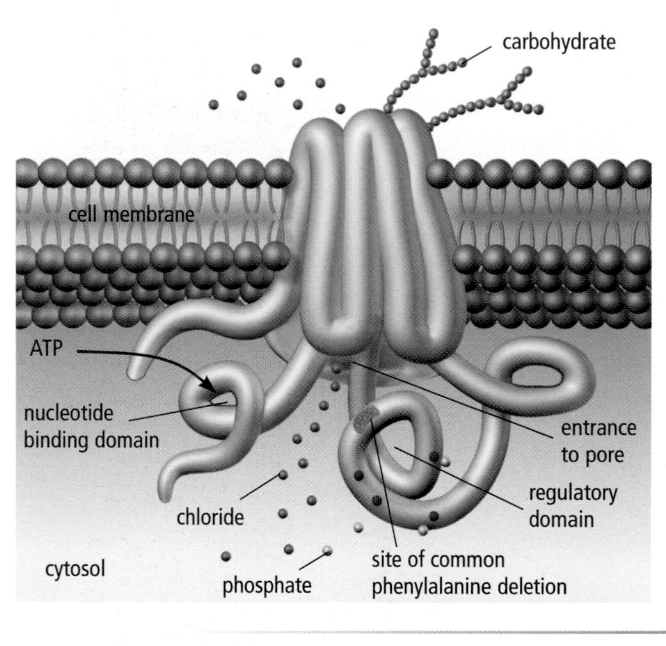

The commonest mutation is a deletion of one base triplet (codon). The mutant protein, therefore, has only 1479 amino acids and cannot fold properly. The mutant form does not channel Cl^-, so the cells become more hypertonic. Water is then drawn into the cells by osmosis, which in turn causes the mucus lining the lungs and gut to become sticky. The cilia in the trachea cannot move this sticky mucus effectively. Regular physiotherapy and drug use is required to alleviate symptoms and reduce the chances of secondary infections. It is estimated that a mutant CF allele is carried by 1 in 20 of the UK population, and this results in the condition being present in about 1 in 2000 births. The high prevalence of this mutation in the population may well be due to a historic selective advantage that carriers gained – some scientists have suggested a heterozygous advantage in the form of protection from cholera.

Function of the normal CFTR protein. The normal CFTR protein is a chloride channel that opens when ATP binds to both nucleotide-binding domains. The ability to channel chloride ions is lost in the mutant forms of the protein that cause cystic fibrosis.

DUCHENNE'S MUSCULAR DYSTROPHY (DMD)

Duchenne's muscular dystrophy is a sex-linked recessive condition – the gene is on the X-chromosome and not on the Y-chromosome. For this reason, the condition is much more commonly seen in males (XY), as they only have one copy of the gene and any mutation will be expressed in their phenotype.

If you use YouTube to watch some first-hand accounts of life with either CF or DMD you will realise why so much research effort is being put into finding cures for these genetic diseases.

The non-mutant form of the gene is activated during puberty and is responsible for the synthesis of the body's largest protein, dystrophin. Dystrophin has 3865 amino acids and links the cytoskeleton to the membrane of muscle cells. The gene, of course, is correspondingly large, and is prone to deletion mutations. Approximately 1 in 3500 of newborn males will develop Duchenne's muscular dystrophy. The condition causes muscle deterioration during adolescence and, currently, is incurable.

contd

SCREENING FOR GENETIC DISORDERS

Genetic disorders can be detected through the use of gene probes. Before this type of test can be developed, the defective gene must be identified by mapping and sequencing. In addition, this is often backed up by an understanding of the normal and mutant protein structures and functions. Once individuals who are carriers or who are at risk of developing disease are identified by screening, appropriate counselling is essential. Counselling usually focuses on clear explanations of likely risks and outcomes, and covers possible treatments and options.

DON'T FORGET

Short, complementary, single-stranded probe sequences are designed to hybridise with specific sequences, genes or gene mutations.

Probing for mutations

The screening for genetic disorders relies mainly on the use of single locus probes (see page 35). In some screening tests (such as for cystic fibrosis), common mutations are identified using probes that hybridise with particular mutant gene sequences. In other screening tests (such as for Duchenne's muscular dystrophy), short sections of the *normal* gene sequence are probed for; any failure of probe hybridisation is a sign of a mutant allele, such as one with a deletion.

GENE THERAPY

Much research is underway to develop ways to use DNA technology to combat genetic disease. **Gene therapy** involves the replacement of a faulty gene with a working copy of the affected gene. In the case of recessive mutations it is not necessary to remove the defective gene – the insertion of the extra gene is enough.

There are many difficulties with gene therapy. In particular, an appropriate vector has to be found to take the gene into the target cell. Typically, the types of vectors under trial are attenuated (weakened) viruses or liposomes. Both of these are able to penetrate human cell membranes. Side effects can be complex and success rates are low. With viral vectors one risk has been cancer; random insertion of the DNA into the patient's genome can disrupt the genes controlling the cell cycle (see page 7).

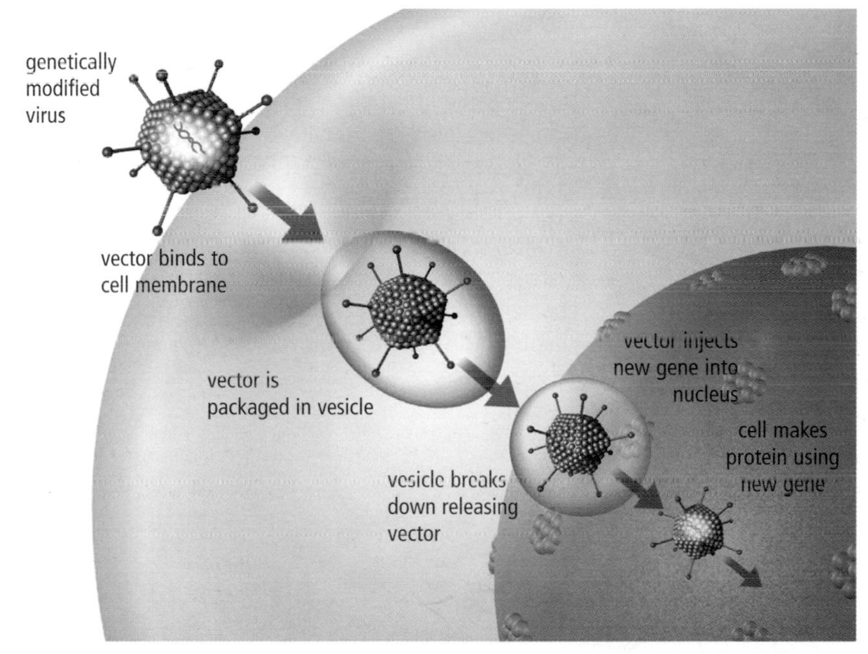

Use of a viral vector in gene therapy

For cystic fibrosis, gene therapy trials have been carried out using a viral vector that is inhaled into the lungs. For Duchenne's muscular dystrophy, attempts are being made to use modified multipotent stem cells that are injected into the bloodstream rather than into muscular tissue.

LET'S THINK ABOUT THIS

The therapeutic use of DNA technology is an area that requires the careful navigation of legal, moral, ethical, medical and scientific issues. There is always a tension between what is possible and what is desirable, and between the needs and autonomy of individuals, and the needs and desires of state or society. For example, is it right to select an embryo created by *in vitro* fertilization on the basis that it will have the correct genotype to provide an organ for transplant into a sibling? For further insight into such issues look at the websites of The Nuffield Foundation or ethics.sandiego.edu.

FORENSIC USES OF DNA

Forensic profiling is the gathering of information to be used in court as evidence. Although DNA profiling is only one aspect of forensic profiling, it is increasingly becoming one of the most important strands in complex legal disputes involving people, animals or other material of biological origin.

DNA PROFILING

DNA profiling, which is also known as genetic fingerprinting, is used to establish beyond reasonable doubt the differences between the genomes of individuals. It is used for forensics, paternity tests, pedigree tests and in cases where the origin of biological material must be determined.

Identifying individuals using DNA

DNA profiling must be able to distinguish individuals from even their closest relatives, so the most variable DNA in the genome must be sought out and compared – we do share about 99.99% of our human genome sequence after all!

Not all of the human genome is made up of genes. In between the genes are regions known as 'junk' DNA, which do not code for proteins. Even within our genes there are non-coding sequences known as **introns** which are removed from the mRNA before it is translated by a ribosome.

Our **exon** gene sequences hardly vary from individual to individual; natural selection tends to allow only the fittest alleles to survive through evolutionary time. The 'junk' DNA, on the other hand, is not influenced greatly by natural selection and sequence changes in these regions have built up steadily through time. This results in the types of variable sequences that can be used to identify individuals.

The most variable (or hypervariable) DNA is located in the short repeating sequences found in areas known as microsatellites or **variable number tandem repeats** (VNTRs). These are repeated short sequences of 3, 4 or 5 nucleotides. The number of repeats is variable, and is inherited. Most of our chromosomes carry sections of VNTRs and they are useful forensically.

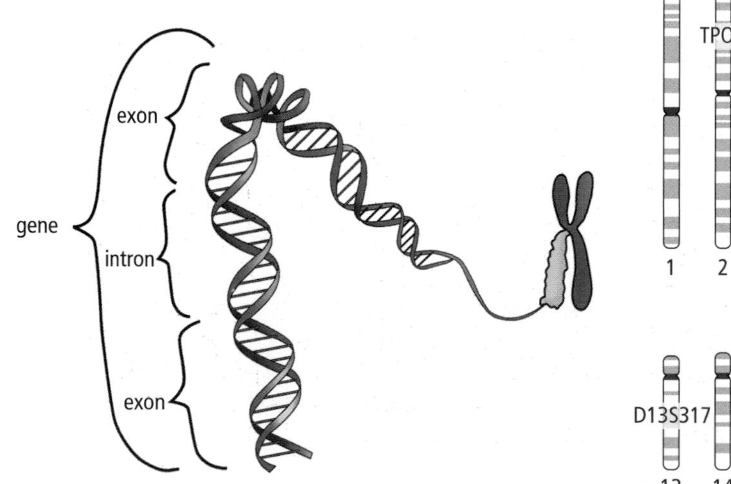

Only the exons of a gene end up in the mRNA used in translation

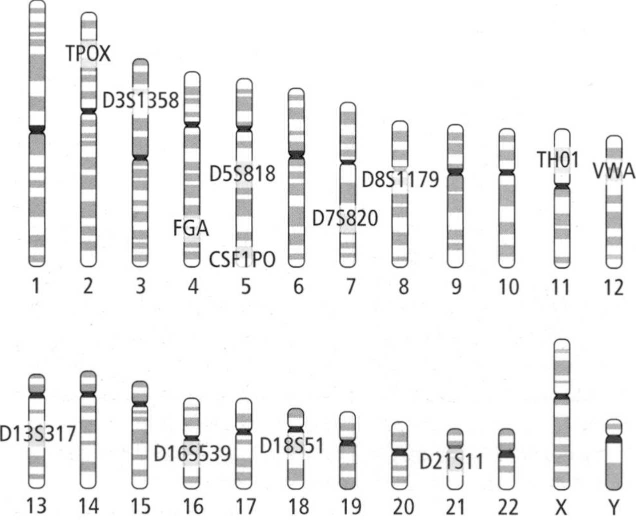

The locations of the 13 short tandem repeat regions (or VNTRs) used in DNA profiling in the US

As they are parts of chromosomes, VNTRs are inherited from both parents. Since they are hypervariable, it is likely that individuals will be heterozygous for the number of repeating units in any particular VNTR. If several VNTRs are examined simultaneously, individuals can be identified, and family relationships can be confirmed by the proportion of shared identical VNTRs.

contd

DNA PROFILING contd

The diagram shows six different numbers of tandem repeating units in one VNTR that produce six different lengths of fragment when digested by a specific restriction enzyme. The different fragment lengths can be identified by probing after separation by gel electrophoresis.

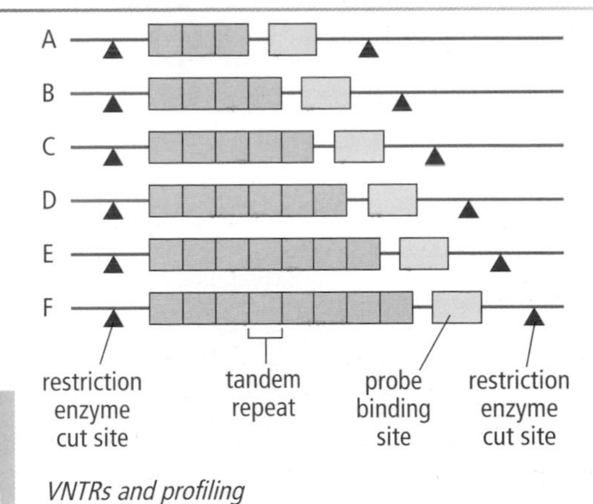

VNTRs and profiling

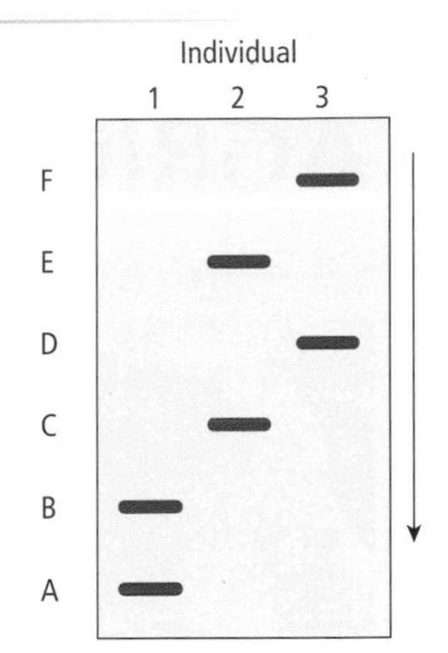

A probed blot for three individuals for this VNTR. Individuals 1 and 2 cannot be the parents of individual 3.

For more information about this work, go to www.dnai.org > applications > human identification

DON'T FORGET

Do not confuse DNA profiling with DNA sequencing! They are quite different.

Stages involved in forensic profiling

To prepare a DNA profile, the following stages must be carried out:

- DNA samples are isolated and may be amplified using PCR.
- The individual DNA samples are digested with a restriction enzyme.
- This enzyme cuts the DNA at specific sequences around the VNTR sites. Since the length of the VNTR is determined by the number of repeating units present in the individuals under test, the length of the fragments produced by restriction digest will differ between different individual genomes.
- The DNA fragments are separated by gel electrophoresis by passing an electric current through the gel. The DNA has a negative charge and shorter fragments travel further in the gel.
- The DNA is denatured into single strands (by heat or sodium hydroxide), and the fragments blotted on to a membrane or filter. A selection of single locus probes hybridises with complementary sequences on the DNA fragments and can be detected (as the probe is labelled).
- The position of the probes produces a banding pattern or fingerprint. This is then compared with other DNA samples treated in the same way. If bands show up in exactly the same positions, then the DNA is from same source. If half the bands are in the same place, this indicates a sibling–sibling or parent–child relationship.
- It is clearly important that no cross-contamination from the operator or between samples is allowed at any stage.

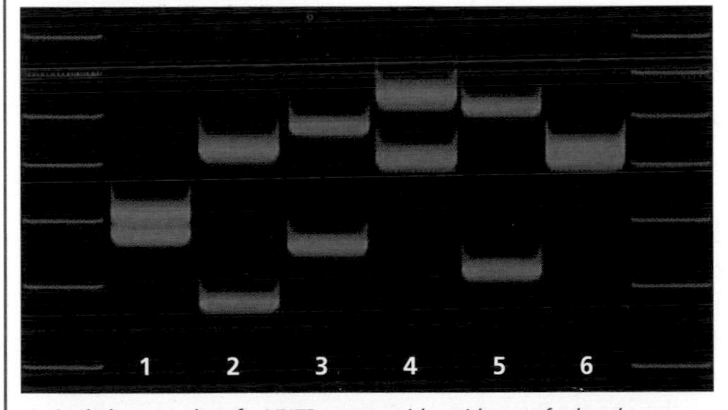

A single locus probe of a VNTR can provide evidence of relatedness. Of the six individuals tested here, numbers 4 and 6 share a band and could, therefore, be siblings or a parent and child.

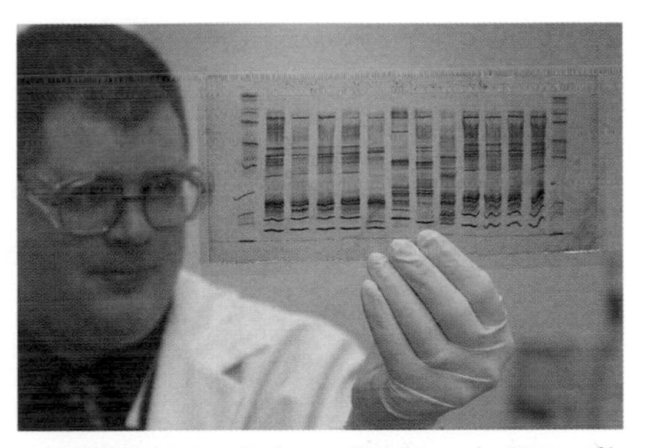

A laboratory analyst checks a multi-locus probe DNA profile to look for a perfect match.

LET'S THINK ABOUT THIS

DNA profiling has become a major tool in forensic science. For example, in 2009 DNA evidence was used to free a man wrongly imprisoned for 27 years. Its use is heavily regulated – the *Human Tissue Act 2004* prohibits private individuals from covertly collecting samples for DNA analysis, but there is a UK database of DNA profiles of criminals kept for reference by the police. There are regular calls for this to be extended to all members of society. What do you think?

DNA TECHNOLOGY IN AGRICULTURE

A Belgian blue bull – a variety produced by selective breeding. This breed has been selected for its 'double muscling'. This is caused by a mutation in the gene for myostatin. The mutation results in a 40% increase in lean muscle growth. As with many breeds, the desirable characteristic comes with a cost: Belgian blue calves are so large that they often have to be delivered by Caesarean section.

Agriculture has a long history of using artificial selection in the modification of species. Selective breeding, for example, has produced countless varieties of domestic animals or crop plants that are suited to human needs.

In **selective breeding**, organisms with certain phenotypic characteristics are chosen for the breeding stock at each generation. Those individuals with unwanted phenotypes are culled or prevented from breeding. In this way, pedigree breeds are maintained through inbreeding. Another technique, **hybridisation**, is the crossing of individuals from different varieties to produce a predictable phenotype in the F_1 generation with dominant characteristics of both breeds.

DON'T FORGET

A *transgenic* organism contains genes from another species within its genome.

GENETIC MODIFICATION

Recent advances in the field of DNA technology have opened several new avenues for artificial selection. The manipulation of genetic information allows genes for favourable traits to be moved from one species to another. The resultant organisms, which contain DNA from two different species, are known as **transgenics**. Some genetic modification involves the artificial transfer of genes between individuals of the same species – such recipients are known as **cisgenics**. An example of cisgenic modification is gene silencing (see page 83).

DON'T FORGET

A bacterium is *transformed* by a recombinant plasmid.

TRANSFORMING BACTERIA

DNA technology is used to genetically modify plasmids so that they will produce eukaryotic protein. To do this an RNA transcript (mRNA) of the desired gene is harvested from a eukaryotic cell and complementary DNA (cDNA) is manufactured using the enzyme reverse transcriptase. This manufactured gene for the protein is then spliced into a plasmid using ligase. Once the plasmid contains the eukaryotic gene, it is known as a recombinant plasmid.

Plasmids and selection genes

A plasmid modified for recombination generally contains a number of important sequences; it has an *ori* region that allows it to be replicated within a bacterium and would usually contain a **marker gene**, such as for antibiotic resistance. The latter allows bacteria that successfully take up the plasmid to be **selected** in a culture – only bacteria with plasmids form colonies on agar containing this antibiotic. In addition, the site for the insertion of the eukaryotic gene is often within another marker gene, such as for resistance to a different antibiotic. Susceptibility to this second antibiotic is used to identify subcultures of bacteria that have successfully taken up the *recombinant* plasmid.

Production of bovine somatotrophin (BST)

An example of the use of genetic transformation of bacteria in agriculture is the production of BST by *E. coli*. The gene was cloned into this bacterial system and the BST protein purified and administered to cattle by injection. BST increases milk production as it prevents mammary cell death and increases the duration of lactation. Consumer concerns about the possible risks associated with the use of this recombinant growth hormone led to it being a banned treatment in many countries. Even in the US, demand is now high for milk to be free of recombinant BST.

TRANSGENIC PLANTS

Transgenic plants can be engineered by either introducing DNA directly into cells using a particle gun or by using the bacterium *Agrobacterium tumefaciens*. The bacterium contains a plasmid that can integrate genes into a dicotyledonous plant genome and can, therefore, be used as a vector in the transfer of DNA from one organism to another. Once in the genome, the transferred DNA is replicated, expressed and passed on like any other gene.

Genetically modified Ti plasmid

The plasmid in *Agrobacterium* is called the tumor-inducing plasmid (Ti). In normal circumstances, the plasmid causes a tumour or gall to form, which increases the habitat for the bacterium.

For genetic transformation, the disease-causing gene is disabled. An additional *ori* sequence is added to allow the plasmid to be grown in the laboratory in *E. coli*. Marker genes are added, such as resistance to the powerful antibiotic kanamycin. The desired gene for transfer is also added. To do this its sequence has to be identified and synthesised, or removed from its source using a restriction enzyme. The plasmid must be cut with the same restriction enzyme and ligase used to seal the gene into the plasmid. The recombinant plasmid can then be returned to *Agrobacterium*.

Cells at the edges of leaf cuttings have been transfected using Agrobacterium *and are forming callus tissue prior to differentiation into genetically modified plantlets.*

Transfection of plant cells

The *Agrobacterium* with the recombinant plasmid can either be directly applied to wounded plant tissue or incubated with plant protoplasts in the laboratory. Protoplasts are prepared by digesting the plant cell walls with cellulase. The plant cell protoplasts are then incubated with the *Agrobacterium* in a **selective medium** (for example containing kanamycin). This allows only those plant cells that have taken up the recombinant plasmid to grow.

Uses of transgenic plants

There are now many examples of potential transgenic crop plants, but few that are grown in the UK. It seems that the general public still has to be persuaded of the benefits of the technology and, at the same time, convinced that it is as low risk as its proponents suggest.

Elsewhere, several genetic modifications have proved successful. For example, a gene for an insect-specific toxin (*Bt*) has been transferred from bacteria to tomato plants, resulting in effective protection from insect damage. There have been trials of transgenic herbicide resistance genes in plants, but these have often been dogged with worries about the 'sideways' transfer of these genes from the crop plant into pest plants via pollen or virus vectors.

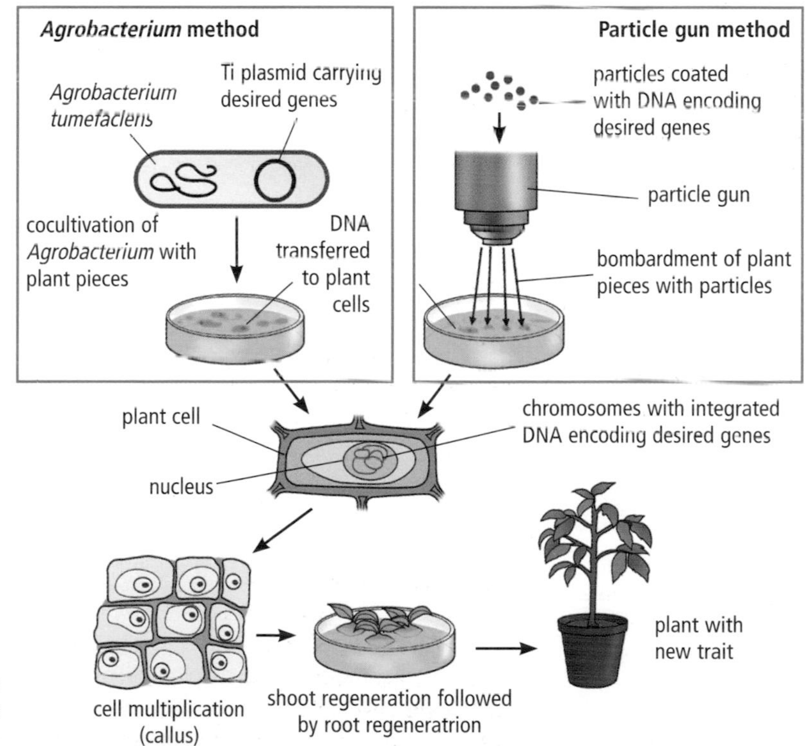

Use of the Ti plasmid vector and a particle gun to genetically modify plant cells

LET'S THINK ABOUT THIS

Strong feelings surround the debate about genetically modified organisms (GMOs). As a student of science, it is important to evaluate the evidence and to consider the validity of trials that have been carried out. To get a feeling for how difficult this is, try reading a few articles about 'golden rice' on the internet and see whether you agree with the genetic modification of rice to contain β-carotene.

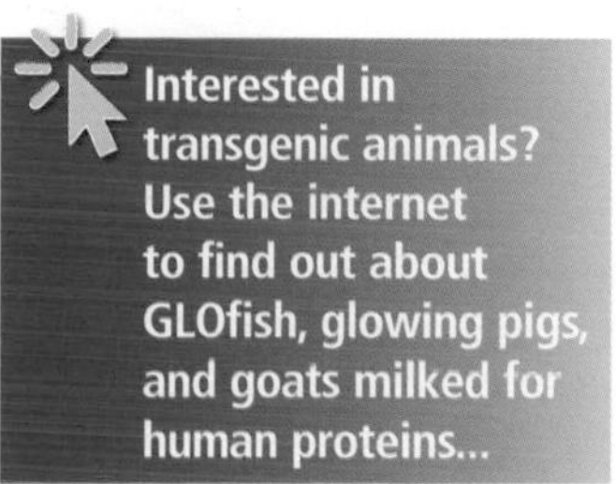

ENERGY FIXATION AND FLOW

ENERGY FIXATION IN ECOSYSTEMS

Energy conversion in organisms

Organisms must convert energy from one form to another to stay alive. In organisms, every energy conversion results in some energy being **lost as heat**. This means that every organism depends on an input of energy and, because an ecosystem has many organisms, the ecosystem itself depends on a **constant input of energy**.

Energy fixation

100 units of energy absorbed by chloroplasts

77 units of energy lost as heat in photosynthesis reactions

23 units of energy built into glucose

9 units of energy used in respiration

14 units of energy built into plant biomass

The energy input for almost all ecosystems comes from the Sun in the form of **solar radiation**, more commonly known as light energy. Some of this light energy can be used by green plants for photosynthesis. Plants are able to fix some of the light energy into chemical energy in organic molecules. These organic molecules are mainly carbohydrates (cellulose and starch), but lipids, proteins and nucleic acids are also made. All the organic molecules together make up the **biomass** of the plant.

Only a small proportion of the light energy absorbed in photosynthesis actually ends up in biomass. A lot of the energy is lost as heat before the final product is made.

Primary productivity

The rate of accumulation of biomass by plants can be measured and is called the primary productivity. The **gross primary productivity** (GPP) is the total yield of organic matter from photosynthesis. However, some of this organic matter is used for respiration to keep the plant alive. So, that leaves the **net primary productivity** (NPP) which is the biomass remaining after respiration.

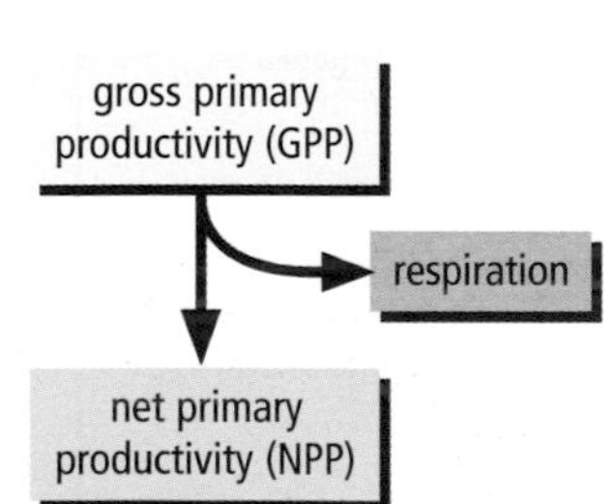

How energy is lost before it forms biomass in plants.

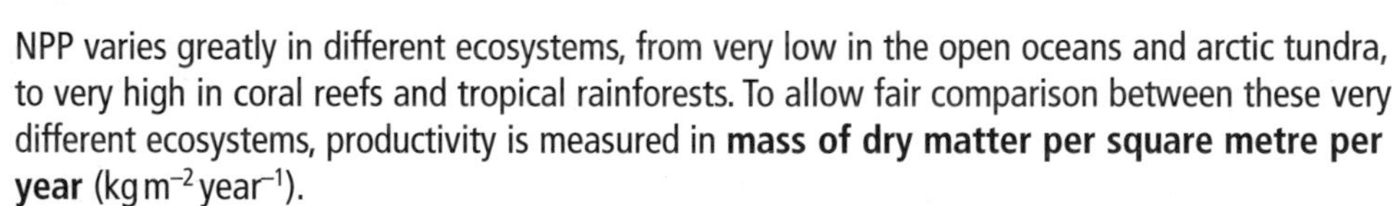

NPP varies greatly in different ecosystems, from very low in the open oceans and arctic tundra, to very high in coral reefs and tropical rainforests. To allow fair comparison between these very different ecosystems, productivity is measured in **mass of dry matter per square metre per year** (kg m^{-2} $year^{-1}$).

NPP represents the plant organic matter that is available to herbivores to use as food. It follows that NPP determines how much productivity of organic matter is possible by animals. Arctic tundra, with its very low NPP, supports much less animal biomass than is found in tropical rainforests.

DON'T FORGET

Photosynthesis traps light energy and converts some of it into the chemical energy stored in organic molecules.

ENERGY FLOW IN ECOSYSTEMS

Energy transfer

Plants use light energy and simple inorganic molecules to make complex molecules to keep themselves alive. All other organisms need energy and nutrients to stay alive and obtain these by consuming complex organic molecules. So, the energy of the plants is transferred through different organisms in the ecosystem by **feeding**. The energy transfer pathway can be drawn out as a food web, with the arrows showing the direction of energy transfer. The food web can be thought of as having different **trophic levels** occupied by organisms that share the same position in food chains.

DON'T FORGET

Some of the GPP has to be used in respiration to keep the plant alive. NPP is what is left.

Trophic levels

The first trophic level (T1) is the **producers**; they are able to use inorganic molecules to form complex organic molecules. These organisms are also known as **autotrophs** and they are most commonly photosynthetic (photoautotrophs).

contd

ENERGY FLOW IN ECOSYSTEMS contd

All other organisms are **heterotrophs**. This term refers to any organism which has to gain energy and nutrients by consuming complex organic molecules. These organisms are also known as **consumers**.

The primary consumers are **herbivores** (T2) because they feed on plants. Carnivores (T3 and above) get their energy and nutrients by eating other animals. Some animals feed on both plants and animals, so they are called **omnivores**.

Another group of heterotrophs gets its energy by feeding on dead organisms or waste material (detritus). These are the **saprotrophs** ('feeding on rot'), which can be divided into two groups. First, the **detritivores** – invertebrates such as the earthworm – which eat pieces of the material and, second, the **decomposers** (bacteria and fungi) which secrete enzymes onto the material to digest it. The saprotrophs are very important in nutrient cycling.

DON'T FORGET

Don't confuse the terms *heterotroph* and *omnivore*!

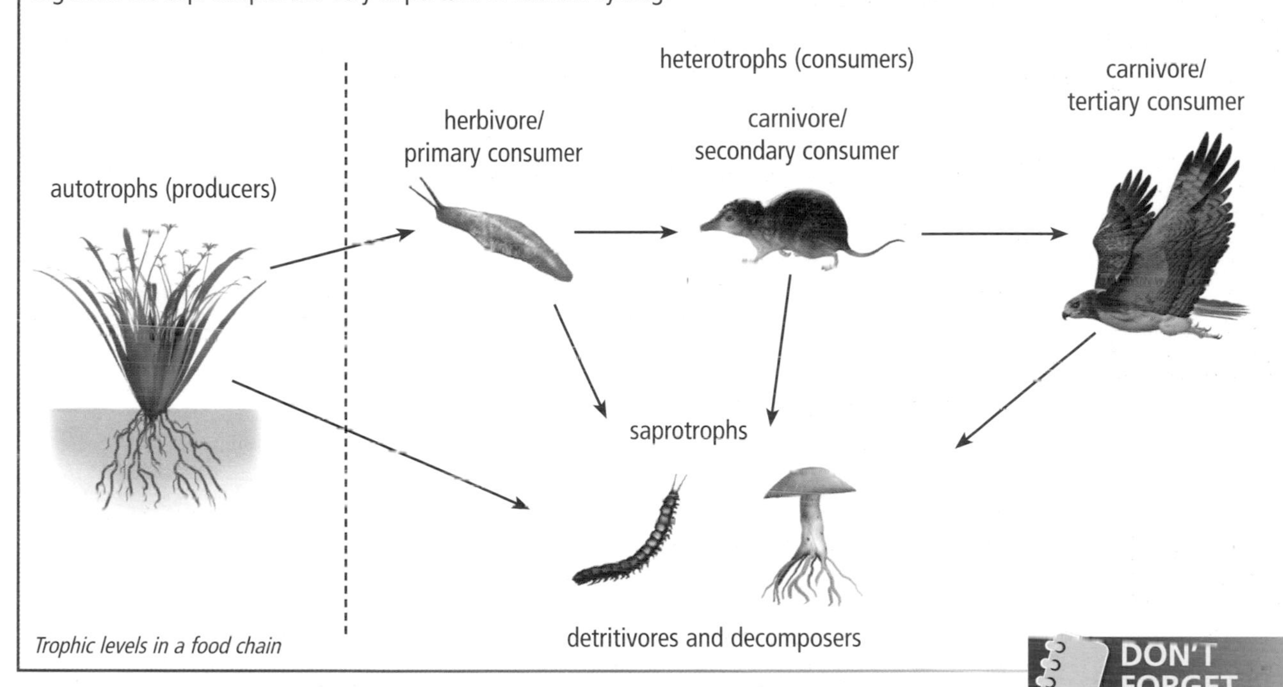

Trophic levels in a food chain

DON'T FORGET

Detritivores take in the material and digest it using enzymes inside their body – they have **internal enzymatic digestion**. The decomposers secrete enzymes on to the food and digest it outside their cells – so they have **external enzymatic digestion**.

As the world population increases, will NPP be able to keep pace with human requirements? Look up http://nasadaacs.eos.nasa.gov/articles/2007/2007_plants.html to see how NASA data about NPP is being used to study this problem.

LET'S THINK ABOUT THIS

Plants are normally producers. There are a few plants which have no chlorophyll and have become consumers. Yellow bird's nest (*Monotropa hypopitys*) is even more bizarre; it has become a saprotrophic plant feeding on leaf litter. It is clearly a plant as it has scale-like leaves and yellow flowers, but it gets food in the same way as a fungus!

Yellow bird's nest – a plant that is not a producer.

ECOLOGICAL EFFICIENCY

ENERGY USE AND LOSS BY ORGANISMS

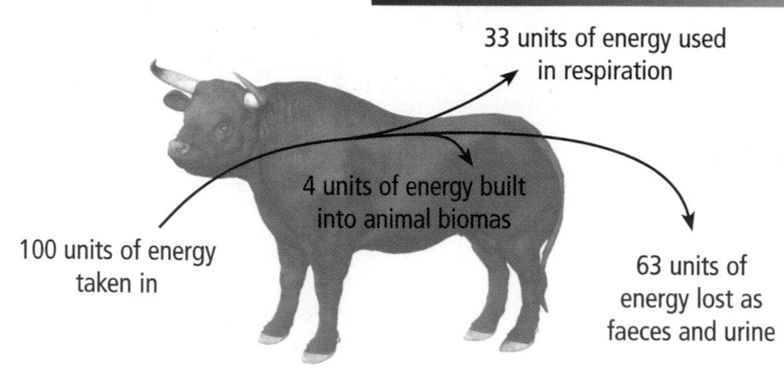

Secondary production in a bullock

Animals use the energy from their food to carry out the processes essential to keep them alive. Some of the energy is used to make organic materials for the growth of the animal – this is called **secondary production**. However, only a little of the food that is eaten is converted into biomass. Some of the ingested food is **undigestable**, so it passes through the digestive system to be egested as **faeces**. Food material that does get digested and then absorbed is used mainly for **respiration**, so a great deal of energy is lost from the animal as **heat** and as **movement**.

DON'T FORGET

Only the biomass of the animal that is made by secondary production will be available to the next trophic level.

Some animals are more **efficient** at converting their food intake into growth. If you study the table below you will notice that the **homeothermic** animals (birds and mammals) use a lot of energy to regulate their body temperature and so have less energy available for secondary production. **Poikilotherms** (such as invertebrates and fish) have more energy available for growth as they do not use energy from their metabolism to regulate body temperature.

Temperature regulation	Body temperature	Animal	Feeding preference	Percentage of energy intake built into animal biomass
poikilothermic	conforms to environment	grasshopper	herbivore	13
		perch fish	carnivore	22·5
homeothermic	regulated	cow	herbivore	1
		owl	carnivore	less than 0·5

DON'T FORGET

A major reason for low energy transfer values is the heat loss associated with maintaining a high body temperature.

MEASURING ECOLOGICAL EFFICIENCY

Food webs are complex

It would be very difficult to work out the ecological efficiency for each species in a food web. The energy flow in food webs is **immensely complex** because most animals feed on a very wide range of food sources. If a food web was to be drawn accurately, there would be so many arrows representing energy flow that the diagram would become completely unreadable! It is more feasible to study how much energy is transferred between the trophic levels.

Energy transfer between trophic levels

There are **significant losses of energy** at each transfer from one trophic level to the next. Only a small fraction of energy that is taken in by an animal becomes secondary production; most of the energy is lost as heat and movement or was simply not available because the food could not be digested. The term **ecological efficiency** describes the **proportion of energy** taken in by one trophic level that is converted into new biomass and so is available to the next trophic level.

Some of the food that is eaten does not get used by the animal. Undigested materials pass into the faeces. There is still a lot of organic material available in faeces that the detritivores and decomposers (saprotrophs) use as a food source. This results in energy flow through **saprotrophic food chains**. Dead organisms are also a food source for the saprotrophs. The saprotrophs are able to use the energy for biomass production. Eventually, the energy will be used for respiration and lost as heat and movement.

DON'T FORGET

Because energy is always being lost as it is passed on through the trophic levels, less and less energy is available. Note that the percentage value for efficiency of transfer between producer and consumer, or between consumers, is not fixed, but varies depending on many different factors.

contd

MEASURING ECOLOGICAL EFFICIENCY contd

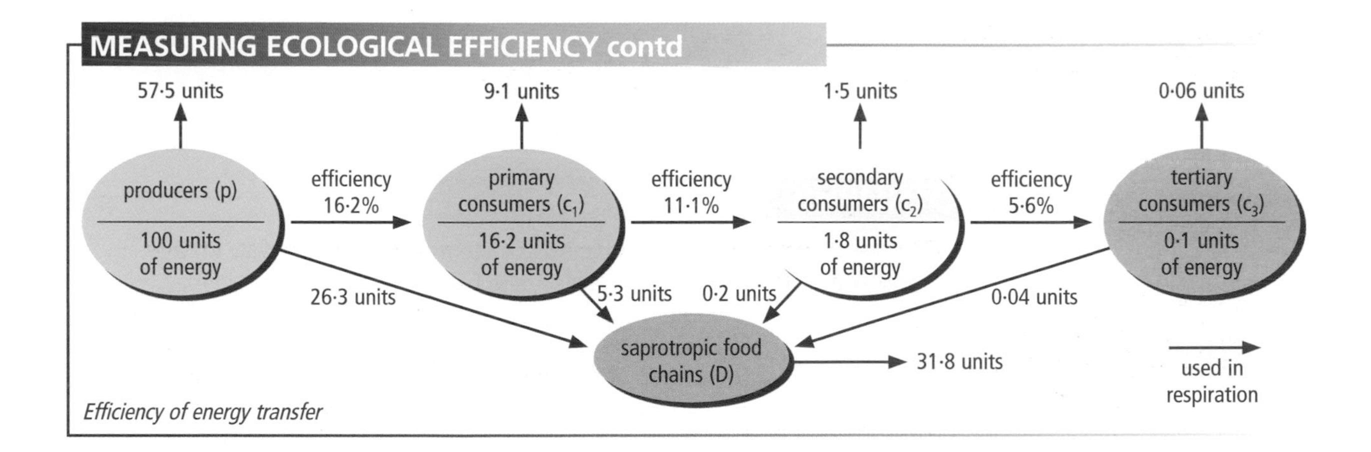

Efficiency of energy transfer

BIOLOGICAL PYRAMIDS

Pyramids of energy flow (pyramids of productivity)

These show **energy built into biomass** at each trophic level in a unit area in a year. The units used are **kJ m^{-2} yr^{-1}**. Each level of the pyramid gets smaller because there is less energy available for production for the next trophic level due to losses as heat and movement.

This type of pyramid gives an accurate representation of ecological efficiency because it decribes the energy built into the organisms and takes account of changes in the productivity at different times of year. But it is very difficult to measure: the organisms have to be burnt to assess the energy content; and the data have to be collected for a full year!

67 C_3
21 × 10^3
1602 C_2
14 × 10^3 C_1
87 × 10^3 P

American river (productivity in kJ m^{-2} yr^{-1})

Pyramids of numbers

These show the **total number of organisms** at each trophic level. They are much simpler than pyramids of energy flow as the sampling of organisms does not take place over the whole year. But there are clear limitations: if the count is made in summer, generally there will be more animals present than in winter; and no account is taken of the size of the organisms, so including a large plant gives an unbalanced pyramid, as shown in the oak forest pyramid.

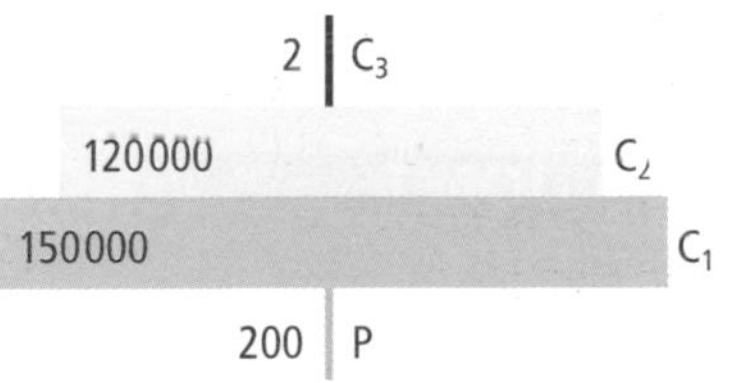

English oak forest (number of individuals per 1000 m^2)

Pyramids of biomass

These show the **total dry mass of organisms** at each trophic level. This type of pyramid is more accurate than a pyramid of numbers because it allows for the size of the organisms shown, but it does not take account of the energy content of the biomass. Fat has twice as much energy per gram as protein or carbohydrate, so a fatty animal would be under-represented! It also does not allow for changes in biomass present over the year. In a marine ecosystem (such as the English Channel), an unbalanced pyramid can result when the rapid reproduction of producers supports longer-lived consumers. The plant plankton keep replacing every two days while the consumer animals live on for weeks or months.

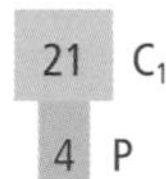

English Channel (biomass in g m^{-2} of water surface)

DON'T FORGET

Pyramid diagrams can be used to represent the ecological efficiency of the ecosystem

Look at http://users.rcn.com/jkimball.ma.ultranet/BiologyPages/F/FoodChains.html to read more about this topic.

LET'S THINK ABOUT THIS

Some mammal carnivores are very elusive and difficult to count. Perhaps ecological efficiency could be used to estimate population sizes? A survey of 25 different types of mammal carnivore (from a weasel to a polar bear) found that 10 000 kg of prey will support only 90 kg of carnivore. If we estimate the biomass of prey animals and we know the average mass of the carnivore, we can estimate how many carnivores may be present in the ecosystem.

DON'T FORGET

Pyramids of number and pyramids of biomass are indirect methods of representing energy flow, so they do not always represent the true pattern of productivity found in ecosystems.

DECOMPOSERS AND DETRITIVORES

SAPROTROPHIC FOOD CHAINS

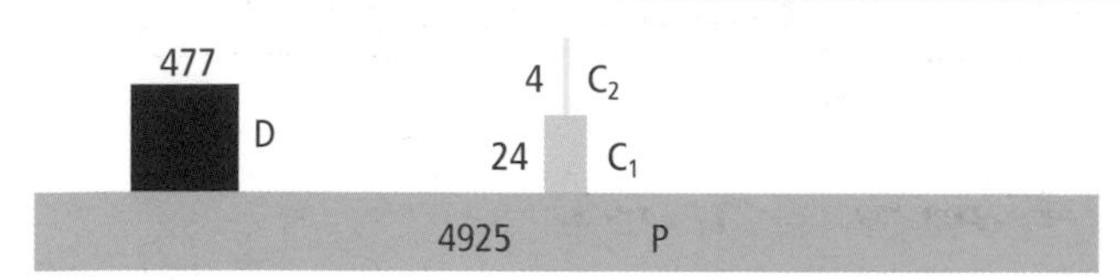

Canadian arctic tundra (productivity in $kJ\,m^{-2}year^{-1}$)

In school, studies of ecosystems tend to concentrate on the grazing food chains. These have familiar organisms in the 'producer → herbivore → carnivore' sequence. In most ecosystems, however, the major food chain is the **saprotrophic food chain**, from detritivores to decomposers. In the arctic tundra pyramid of productivity shown here, notice that box D (the productivity of the detritivores in the saprotrophic food chain) is much greater than the herbivore productivity (C_1 – first consumers).

The detritivores

DON'T FORGET

Make sure you understand the role of fragmentation in helping decomposition.

These are **invertebrates** that eat dead organisms or waste material (detritus). Examples of detritivores include the earthworm, the millipede and the woodlouse. These animals take detritus into their digestive systems. This detritus is then **digested internally** using enzymes secreted into the gut. However, some of the detritus is not digested. As it passes through the digestive system, the undigested detritus becomes **fragmented** into smaller pieces. This passes out as faeces and forms **humus**. This fragmentation increases the surface area of the material available for decomposers to work on.

The decomposers

Decomposition is the chemical breakdown of organic matter in detritus. The organisms which do this are **fungi** and **bacteria**, collectively called the **decomposers**. These are by far the most numerous and diverse organisms in an ecosystem. The fungi are the major decomposers of plant material; they get their energy by breaking down organic matter such as cellulose in dead plants or in herbivore droppings. Bacteria are the major decomposers of dead animals and a wide range of bacterial species will be present using the different molecules available in a carcass.

Decomposition is sometimes called **mineralisation** because nutrients are released into the soil.

Decomposers use **external digestion**. They secrete enzymes out of their cells to break down complex organic compounds into smaller molecules and nutrient ions which can then be absorbed into the cell. Some of these nutrient ions are available in the soil to be absorbed by plants. The decomposers will themselves die and be decomposed, so eventually all the nutrients are recycled into the soil.

The role of **soil microorganisms** is essential for an ecosystem. They are the only organisms able to make mineral nutrients available again for use by plants. Without soil decomposers, dead material would build up and plant growth would cease due to lack of nutrient availability. Nutrients would only be available from the very slow erosion and weathering of rocks!

The process of decomposition is limited by **low temperature** which causes enzymes to work slowly. **Availability of nitrogen** can also limit the rate of decomposition, because the microbes need nitrogen to make enzymes and other cell proteins. These, and other factors, are major considerations during the culturing of microbial cells (pages 10 and 74).

This is the fruiting body of a fungus; the main part of the fungus consists of hyphal threads which are decomposing this mound of rhino dung.

NUTRIENT CYCLING

Nutrients have to be recycled

Energy **flows** through an ecosystem and has an input and an output. The input of energy from sunlight is fixed by producers and this is eventually lost as heat from metabolic processes, mainly respiration.

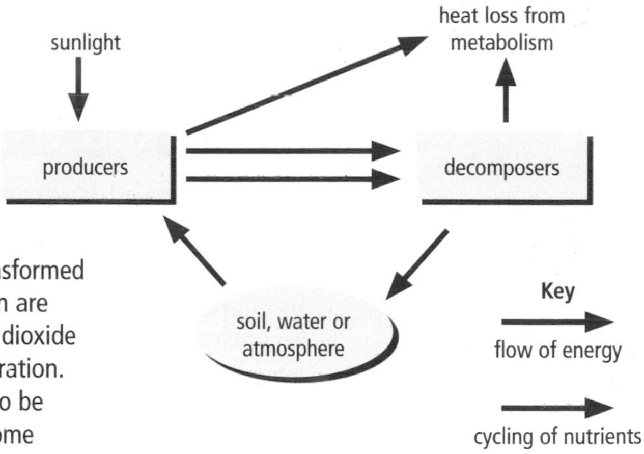

Energy flow and nutrient cycling in the simplest ecosystem possible

Chemical elements, on the other hand, are a finite resource and have to be **recycled**. These **nutrients** are fixed into organisms, transformed into various compounds as they pass along the food chain, and then are broken down into simple forms that can be fixed again. The carbon dioxide fixed by photosynthesis is recycled back to the atmosphere by respiration. All the other nutrient elements are found in compounds that have to be broken down by **decomposer organisms in the soil**, so they become available to the community again.

Nutrient cycling has three key steps

All nutrient cycles have **three key steps** which can be illustrated using the **carbon cycle** as an example.

Stage	Description	An example from the carbon cycle
fixation	the nutrient is taken from the abiotic environment and is fixed into a food web as an organic compound	photosynthesis fixes carbon dioxide to form carbohydrates
transformation	the organic compound can be changed into other molecules	metabolic processes in plants convert the carbohydrates into proteins, lipids and nucleic acids
loss	the organic compounds are lost from the food web; they are broken down to release simple inorganic units back to the abiotic environment	respiration by all organisms releases the carbon dioxide initially fixed by photosynthesis

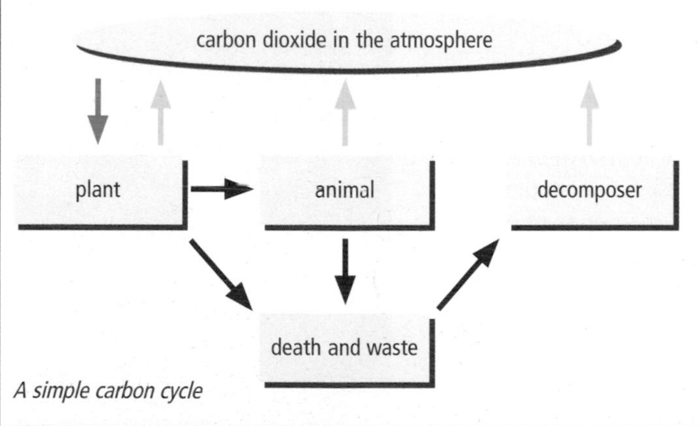

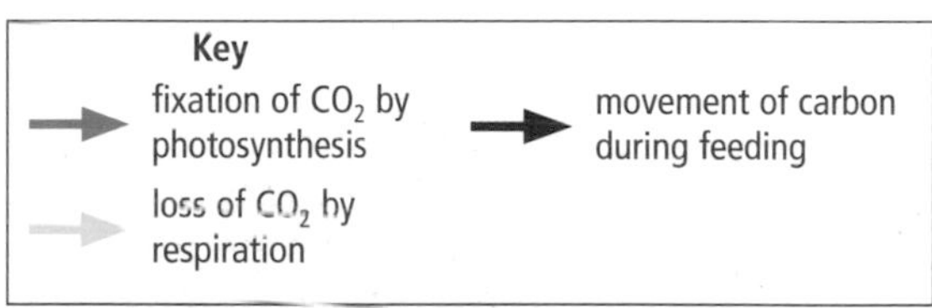

A simple carbon cycle

An aquarium is a tiny ecosystem but, since the aquarist wants lots of fish, food input is needed to supplement the producers. But it's not that easy to replace the detritivores! Look at http://reefkeeping.com/issues/2002-03/rs/index.php for a good explanation of the role of the detritivores in marine aquaria.

LET'S THINK ABOUT THIS

'*If no decay had occurred since the year 1600, dead bodies would cover the Earth to a depth of one kilometre.*' Some textbooks try really hard to persuade us of the importance of decomposers! But, actually, the situation would be even more drastic than this.

If all the decomposers stopped working, there would be no more nutrient cycling. Sufficient carbon dioxide might get back into the air to supply photosynthesis, but all the other nutrient elements would be locked up in the dead materials. Nutrients would not return to the soil to supply plant growth. After the first generation of organisms had died, plant growth would come to a halt within a few years. The 1 km depth of dead bodies would not happen – life would stop sooner than this.

NUTRIENT CYCLES

THE NITROGEN CYCLE

The essential bacteria

Nitrogen is a component of proteins and nucleic acids. Nitrogen makes up about 79% of the air and yet the productivity of many ecosystems is limited by a lack of *available* nitrogen. This is because the nitrogen in the air is present as nitrogen gas (N_2) which is very **inert** – it does not react easily with other chemicals, so it is not accessible to plants and animals. The only way that nitrogen can be made available to the organisms in an ecosystem is by the action of bacteria.

Fixation of nitrogen

Lightning can provide the high energy required to combine atmospheric nitrogen with oxygen from water to form nitrates; only about 2% of the nitrogen fixation on Earth happens this way. About 20% is done industrially by the Haber process and the remaining 78% or so is fixed by just a few bacterial species.

Biological nitrogen fixation uses the **nitrogenase enzyme** which is able to fix **nitrogen into ammonium**. But this enzyme has two big limitations: it needs a lot of energy (16 ATP molecules for every nitrogen molecule fixed) and it is competitively **inhibited by oxygen** (see page 27).

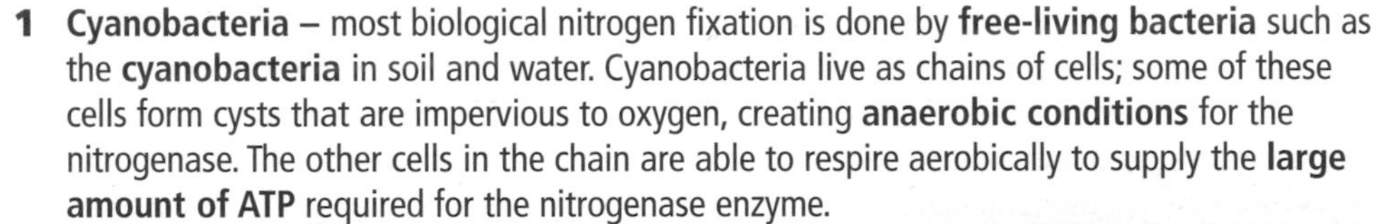

A cyanobacteria chain showing a heterocyst. The heterocyst is about 5 μm wide.

There are two main groups of nitrogen-fixing bacteria and they have evolved different methods of overcoming these two problems.

1. **Cyanobacteria** – most biological nitrogen fixation is done by **free-living bacteria** such as the **cyanobacteria** in soil and water. Cyanobacteria live as chains of cells; some of these cells form cysts that are impervious to oxygen, creating **anaerobic conditions** for the nitrogenase. The other cells in the chain are able to respire aerobically to supply the **large amount of ATP** required for the nitrogenase enzyme.
2. ***Rhizobia*** – about 10% of biological nitrogen fixation is done by various species of the genus ***Rhizobium.*** These form a mutualistic relationship (see page 61) with **legume plants** like peas, beans and clover. *Rhizobia* live in **root nodules** that the plant fills with **leghaemoglobin**. This molecule binds to the oxygen in the nodule to create the **anaerobic conditions** necessary for nitrogenase function. Oxygen is then released slowly for aerobic respiration, so that the bacteria can produce the **large amount of ATP** required by nitrogenase without oxygen acting as an enzyme inhibitor (see page 27).

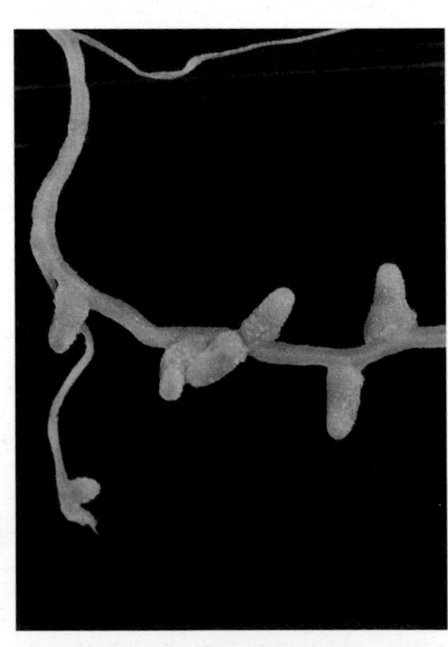

Root nodules look pink because of the presence of leghaemoglobin. Each nodule is about 3 mm across and contains millions of Rhizobia bacteria.

DON'T FORGET

Rhizobia are mutualistic with legumes; cyanobacteria are free-living in soil and water.

Transformation of nitrogen

Fixed nitrogen has to enter plants in order to become available to the rest of the ecosystem. Some of the **ammonium** formed by cyanobacteria diffuses into the soil and is **absorbed by plant roots**. Some of the ammonium is converted to **nitrate** ions before being absorbed by plants. *Rhizobia* convert their ammonium to simple amino acids (see page 17), some of which are absorbed directly into the legume plant. Once inside plants, the nitrogen is **assimilated** through the metabolic reactions that form **proteins and nucleic acids**.

Some of these plant proteins are eaten, so nitrogen is transformed to make animal proteins. When plants and animals die, their proteins are decomposed by bacteria to release ammonium – this is also known as **ammonification** (a mineralisation process; see page 50). Some of this ammonium can be absorbed directly by plants.

contd

THE NITROGEN CYCLE contd

Ammonium not absorbed directly by plant roots **is converted into nitrates** in a process called **nitrification**. Two different types of **nitrifying bacteria** are required:

1. ***Nitrosomonas*** bacteria convert the **ammonium to nitrite ions**.
2. ***Nitrobacter*** convert the **nitrite to nitrate ions** which can then be absorbed by plants.

Both these bacterial types are **chemoautotrophic**, gaining energy by oxidising inorganic ions.

DON'T FORGET

Don't confuse nitrogen fixation (nitrogen to ammonium) with nitrification (ammonium to nitrite to nitrate).

Loss of nitrogen

Nitrogen can **leach** out of the soil ecosystem in water. This may lead to **eutrophication** (see page 69). Another loss happens by **denitrification** as **nitrate is converted to nitrogen gas**. This only occurs if a soil is **water-logged** and **anaerobic**. Free-living bacteria such as ***Pseudomonas*** cause denitrification as they use the oxygen from nitrate ions for respiration.

DON'T FORGET

Denitrifying bacteria are only active in anaerobic conditions.

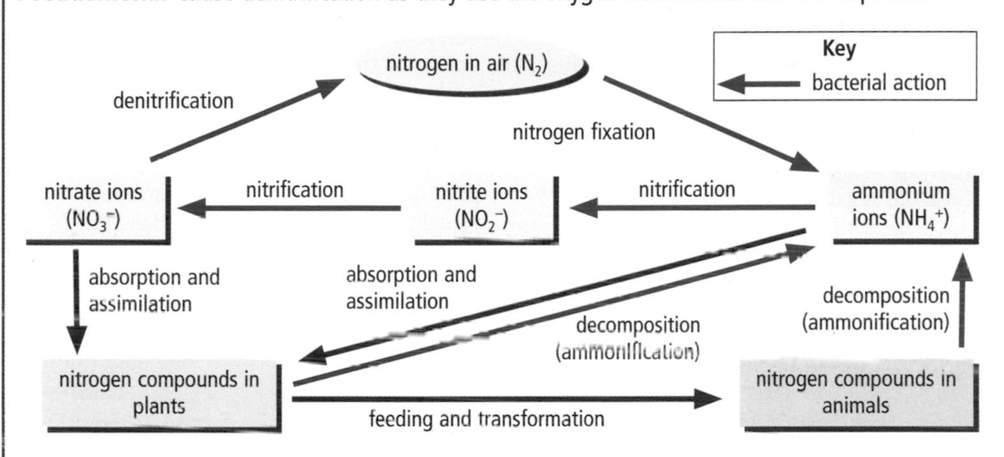

The biological stages in the nitrogen cycle

THE PHOSPHORUS CYCLE

The biological stages in the cycling of phosphorus are very simple. Plants absorb phosphate ions from the soil and fix them during the production of ATP, nucleic acids and phospholipids. Animals get their phosphates by eating plants. Decomposers mineralise phosphates back into the soil from dead organisms.

Phosphates are readily lost from an ecosystem. In land ecosystems, this is partly because the soluble phosphate ions are quickly leached out of the soil, and partly because phosphate ions combine with other chemicals to form insoluble compounds which cannot be used by plants. In **aquatic ecosystems**, phosphate deficiency is a **major limiting factor of productivity** as the insoluble compounds settle to the bottom. Millions of years later, these sediments have become rocks which may be pushed up to the land. Here they can be eroded and the phosphate in them slowly washed back into soil and rivers.

Because phosphate ions are usually in short supply, evolution has selected plants that are able to absorb them quickly. Fertilisers can be used to add phosphates to encourage crop growth. But if the fertiliser leaches from agricultural land into an **aquatic ecosystem** then **phosphate enrichment** may occur; this is also known as eutrophication (see page 69).

This long link is a good animation of the nitrogen cycle showing the movement of nitrogen atoms through an ecosystem. http://www.mhhe.com/biosci/genbio/tlw3/eBridge/Chp29/animations/ch29/1_nitrogen_cycle.swf

LET'S THINK ABOUT THIS

Carnivorous plants, such as the sundew from Britain, are only found in swamps or marshes. Why? Water-logged soil has very low nitrogen content due to denitrification. Carnivorous plants have evolved a way of getting nitrogen by trapping insects and secreting enzymes to digest them.

A sundew getting its nitrogen.

PREDATION AND GRAZING

PREDATION AND GRAZING ARE +/– BIOTIC INTERACTIONS

A herring gull predates a gannet egg, thus killing the developing gannet.

Predators and grazers are **heterotrophs**; they derive their energy and nutrients by **consuming** other organisms or materials of organic origin. This is a **+/– interaction** because the interaction is beneficial (+) to the predator or grazer (through energy gain), and is detrimental (–) to the food species because it causes injury or death.

Predation

Predation causes the **death of the prey** organism. In many cases, the entire prey organism is consumed by the predator. Some ecologists take a wider view and consider any feeding on living animal or plant tissue to be predation. In Advanced Higher Biology, predation refers only to **animals preying on other animals**.

DON'T FORGET

The effect of a density-dependent factor increases as the population density increases. These factors tend to be biotic factors.

Predator–prey cycles

In an ecosystem with low biodiversity and a **simple food web**, there are **strong links** between the individual predator and prey species. This means that any change in the abundance of one species will have a **density-dependent** effect on the other. The lynx and snowshoe hare from North America provide a good example of the resulting unstable (chaotic) cycles of predator and prey abundance. An increase in lynx density leads to a reduction in the density of snowshoe hares. Fewer hares mean lynx density decreases. The lower predator density allows the prey density to increase, which then leads to an increase in lynx density once again.

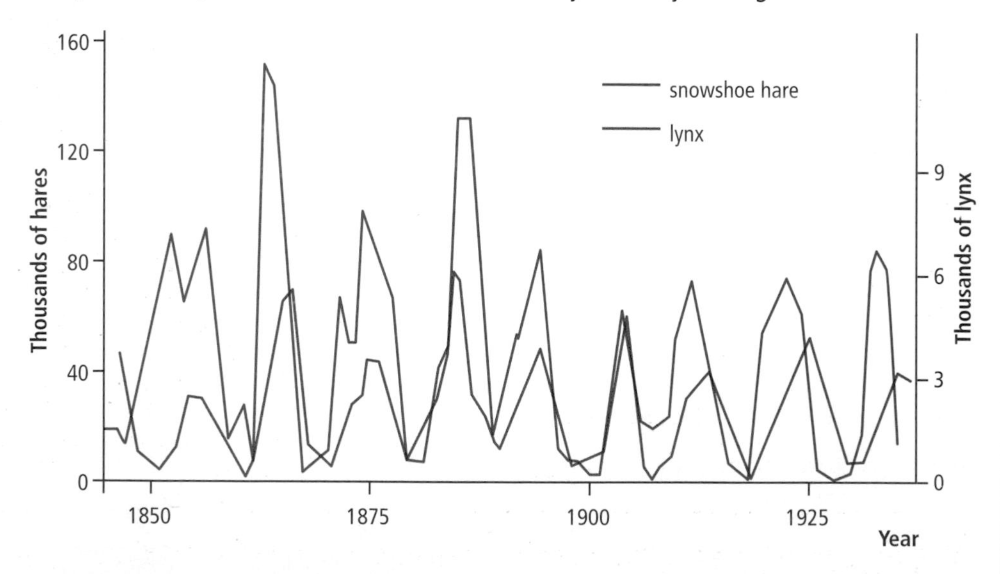

Predator–prey cycle relationship between lynx and snowshoe hare

In most ecosystems, predator–prey cycles are rare. Higher biodiversity and more complex food webs reduce the dependence of individual predator species on individual prey species, so the ecosystem has higher stability. Predators can switch between food sources, so population sizes do not tend to fluctuate wildly.

This website has a long URL but it gives the broader picture of the effect of the snowshoe-hare population on the rest of the community. Read right the way through – it's even more interesting in reality! http://www.wc.adfg.state.ak.us/index.cfm?adfg=wildlife_news.view_article&articles_id=339

contd

PREDATION AND GRAZING ARE +/– BIOTIC INTERACTIONS contd

Grazing

Grazing is usually defined as the **consumption of green plants or algae by herbivorous animals**. In fact, strictly, grazing refers only to grass-eating animals rather than to those that **browse** on broad-leaved plants (such as the bushbuck in the picture). In grazing, generally only part of the plant structure is removed for consumption. As only part of the prey item is removed, **grazing is not usually lethal** and regrowth can occur.

Grazing has provided a strong evolutionary selection pressure for plant species with **basal meristems** (low-growing points), such as grasses. These species can tolerate grazing better than plants with **aerial meristems** (high-growing points).

A bushbuck browses vegetation, causing damage and injury to the plants, but not necessarily death.

- At low-grazing intensity, the plant community is dominated by a small number of competitive grass species.
- At moderate-grazing intensity the more competitive plant species are kept in check, so a wider variety of less vigorous species get a chance to grow as well.
- At high-grazing intensity many species are unable to recover from frequent grazing. Only the few species that are able to cope with the damage from grazing will survive.

DON'T FORGET

Predators kill their animal prey. Grazers just remove parts of plants, which usually survive.

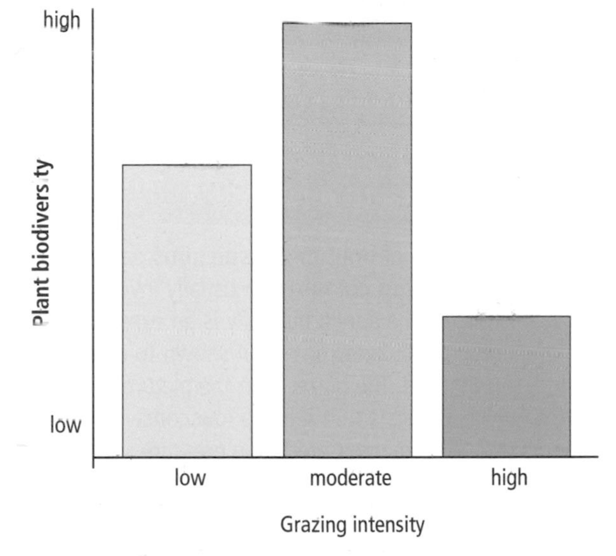

Moderate grazing can increase plant biodiversity

Exclusion experiments reveal the impact of grazing by sheep in upland environments.

Look up http://www.ars.usda.gov/is/AR/archive/may99/plant0599.htm to find out how research on grazing is being used to maximise biodiversity and make cattle ranching more efficient.

LET'S THINK ABOUT THIS

How do predators and grazers shape an ecosystem? Predation and grazing play a very strong 'top-down' role in shaping the structure of the community. The presence of predators and grazers leads to a higher biodiversity. No single species of prey is able to dominate, so weaker competitor species can survive. This increase in biodiversity increases ecosystem complexity, which leads to an ecosystem with higher stability.

DEFENCES AGAINST PREDATION

There is a wide range of defences against predation, including spines, hard shells and poisons. In Advanced Higher Biology we concentrate on various adaptations of colouration for defence. There are thousands of interesting examples – we can only look at few.

CAMOUFLAGE

Camouflage can prevent predators from detecting or recognising prey. In **crypsis**, the prey species blends in with the background, for example the dark arches moth. Like the peppered moth, this is a common species in Britain but it is rarely noticed due to its camouflage. In **disruptive coloration**, the pattern breaks up the outline of the prey so that the body shape cannot be distinguished clearly by predators, as shown by the plains zebra in Africa.

An example of cryptic camouflage – the dark arches moth

An example of disruptive camouflage – the stripes of the plains zebra are thought to confuse predators.

WARNING OR APOSEMATIC COLOURATION

This apollo butterfly has a comparatively small area of orange colouration on its wings.

Aposematic organisms deter predators through the use of **bold or conspicuous** patterns combined with a **distasteful or toxic** nature. The **warning colouration** usually involves bright contrasting patterns of black with yellow or red. The apollo butterfly is an example of an aposematic species in Europe. The extent of its orange markings has been shown to correlate with the concentration of a toxic compound, sarmentosin. The butterfly in the picture has a low ratio of orange areas to total wing area, which suggests that it has a low concentration of sarmentosin; this may be a local evolutionary response to lower predation pressure.

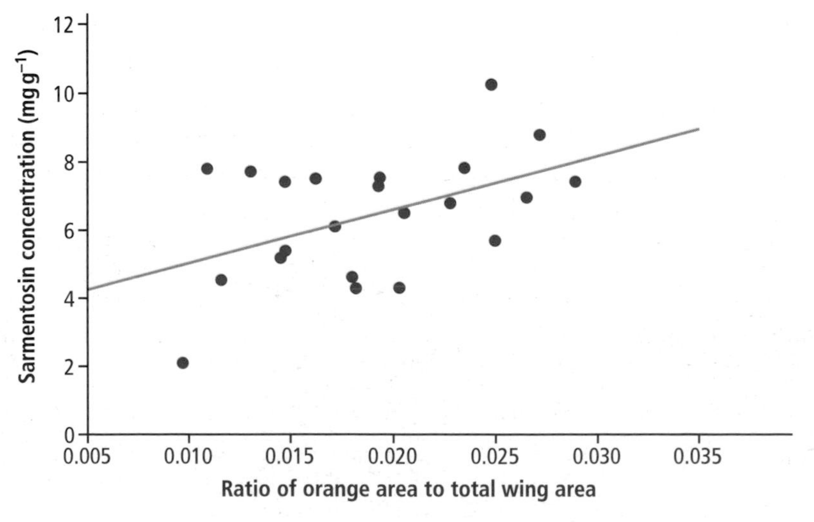

The correlation of colouration and sarmentosin content in apollo butterflies

MIMICRY

Müllerian mimicry

Predators learn to avoid aposematic organisms and, as a result of convergent evolution, many types of toxic aposematic organisms resemble one another. This **resemblance between two harmful** or **unpalatable species** is known as **Müllerian mimicry**. A well-known example of Müllerian mimicry is shown by wasps and bees – both organisms have stings and similar yellow and black colouration.

Müllerian mimicry – the German social wasp and honey bees show similar warning colouration.

Batesian mimicry

Because there are lower levels of predation on aposematic species, **harmless species** that mimic the **appearance of harmful** species can also have increased survival chances. A well-known example of this **Batesian mimicry** is where a harmless hoverfly (the **mimic**) closely resembles a stinging bee or wasp (the **model**).

The evolutionary stable number of mimics in an ecosystem is usually lower than the number of models. If the mimic becomes too common, then the predator will not learn to avoid the aposematic colouration – the prey is usually a harmless food source. The stable number of mimics is also influenced by factors such as the toxicity of the model and the learning efficiency of the predators.

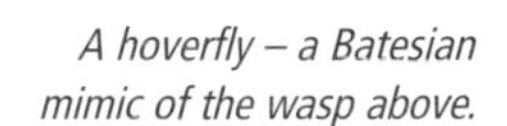

A hoverfly – a Batesian mimic of the wasp above.

DON'T FORGET

'Bates' are fakes.
'Müllers' are real-er!

Aposematic colouration is a very good defence mechanism against predators, but it doesn't guarantee survival. Watch a great bit of film (8.28 minutes) about the biology and survival problems of the golden frog at YouTube by typing *Almost extinct golden frog.*

LET'S THINK ABOUT THIS

In science, knowledge is modified by evidence. For example, when a 'classic' example of Müllerian and Batesian mimicry among butterflies was tested experimentally, it was shown to be false.

The viceroy butterfly had long been considered to be a harmless Batesian mimic of the unpalatable and aposematic Müllerian mimics, the monarch and queen butterflies. Instead, predation rates by birds under experimental conditions showed that the viceroy butterfly is as unpalatable as the monarch. Sometimes it takes time for discoveries like these to reach the pages of textbooks.

The monarch butterfly

The queen butterfly

The viceroy butterfly

COMPETITION

COMPETITION IS A –/– INTERACTION

Competition occurs when two organisms attempt to utilise the same resource. In animals, the resources could be food, water, mates or space. Plants tend to compete for light, water and soil minerals. Competition has a **density-dependent** effect; its impact is greater when population densities are higher, that is, when the resources are shorter in supply. Competition is a **–/– interaction** as it is detrimental (–) to both organisms involved; the costs of competing include reduced growth, reduced fecundity (production of offspring) and increased mortality.

DON'T FORGET

Competition always has a negative impact on both parties – any positive outcome for the winner would have been even greater without the competitor.

Intraspecific and interspecific competition

Intraspecific competition involves members of the same species attempting to use the same resources. **Interspecific competition** involves individuals of different species. Intraspecific competition is said to be **more intense** than interspecific competition because individuals of the same species have identical resource requirements, whereas individuals of different species in the same community may be able to use some different resources. When a population reaches a high density then intraspecific density-dependent competition will be intense and will cause much mortality. On the other hand, introduced exotic species (see page 59) may be successful at interspecific competition with native species.

Interspecific competition between channelled wrack algae and barnacles is for space on upper-shore rock.

Gannets on a rocky colony experience intense intraspecific competition for nest sites.

Exploitation and interference competition

In **exploitation competition**, competitors do not interact directly – one individual uses the resource before the other has the opportunity. This exploitation can occur months before the other competitor is affected. For example, a stronger bite and digestive system allow grey squirrels to exploit under-ripe fruits before they become available to red squirrels.

In **interference competition**, on the other hand, the behaviour of one competitor prevents the other from accessing the resource. For example, the territorial behaviour of many animals is a form of interference competition.

DON'T FORGET

If a resource has been exploited it no longer exists.

NICHE

The concept of **ecological niche** attempts to summarise all those factors that influence the distribution of an individual species. At its simplest level, niche was originally defined as the role an organism plays in the community – such as *woodland herbivore* or *grassland detritivore*. Clearly, however, no herbivore could survive in all woodlands, so a more sophisticated definition of niche is required.

The concept of niche must include many factors (or dimensions). The dimensions that have to be considered include **abiotic factors** (e.g. temperature, humidity, pH, salinity, physical forces, pollution, light intensity, gas concentrations, minerals), as well as the **biotic interactions** within the community (e.g. predator–prey interactions, parasitism). Niche is, therefore, best defined as the **multidimensional summary of an organism's tolerances and requirements**!

contd

NICHE contd

For example, the niche of the barnacle *Semibalanus* is defined by many factors including: substrate (rocky shores), food availability (small particles of food for filter feeding), temperature (found in northern Europe), and its ability to resist desiccation when the tide is out (it is restricted to lower parts of the shore).

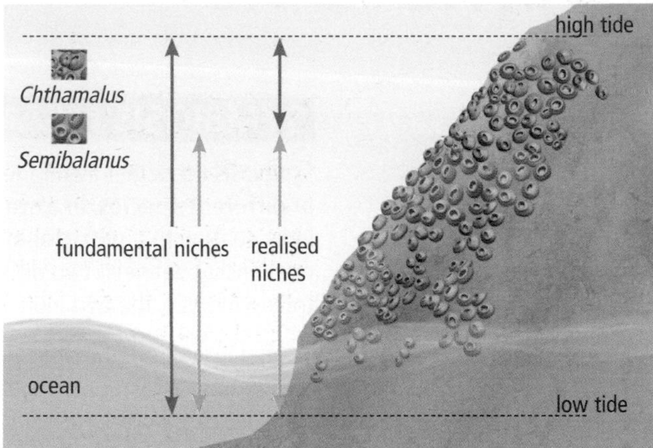

Shoreline zonation – niche at its most visual

Fundamental and realised niche

- An organism's **fundamental niche** is the set of resources that the organism is capable of using in the absence of competition.
- An organism's **realised niche** is the set of resources that it actually uses in the presence of competition.

The fundamental and realised niches of two species of barnacles can be seen in terms of their positions on the intertidal shoreline. The fundamental niche of *Chthamalus* is very wide and extends from low- to high-tide line. The fundamental niche of *Semibalanus* is the lower part of the shore only. When the two species are in competition, the realised niche of *Chthamalus* becomes restricted to the upper shore.

DON'T FORGET

Fundamental niche is the *theoretical* niche, whereas realised niche is the *reality*.

Competitive exclusion principle

Two different species with the **same realised niche cannot coexist** in a community for any length of time. The population of one species will tend to die out within several generations, and this observation is known as the **competitive exclusion principle**.

Resource partitioning

Species can coexist in similar niches if **resource partitioning** evolves. When different species have similar requirements natural selection will lead to one or both species exploiting different components of the resource, thus reducing competition. The evolution of different beak lengths and foraging behaviours in waders has allowed various species to exploit food of different types and at different depths in their shared habitat.

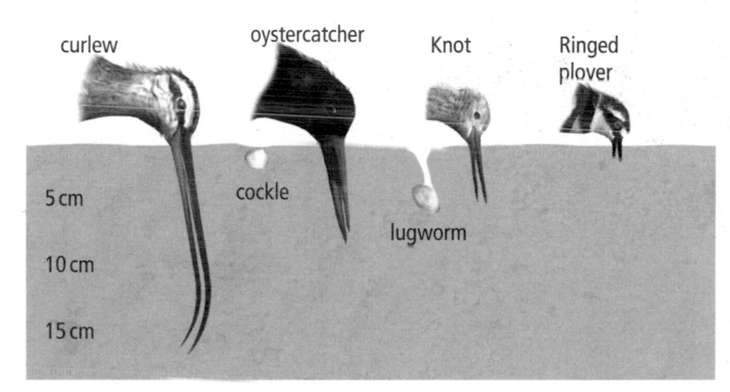

The beaks and foraging depths of wading birds

Impact of exotic species

Exotic species, also known as foreign or alien species, are those that have been introduced into a community as a **result of human activity**. Some exotic species have a **competitive advantage** over native organisms (for example, a lack of local predators or pathogens), whereas others are able to fill an unoccupied niche.

Population explosions of exotic species can cause damaging effects in terms of population decrease of native species and **loss of diversity**. It is much easier to introduce an exotic species into an environment than to eliminate a damaging one. Examples in parts of the UK include giant hogweed, rhododendron, Japanese knotweed, Himalayan balsam, signal crayfish, grey squirrel, hedgehog and ruddy duck.

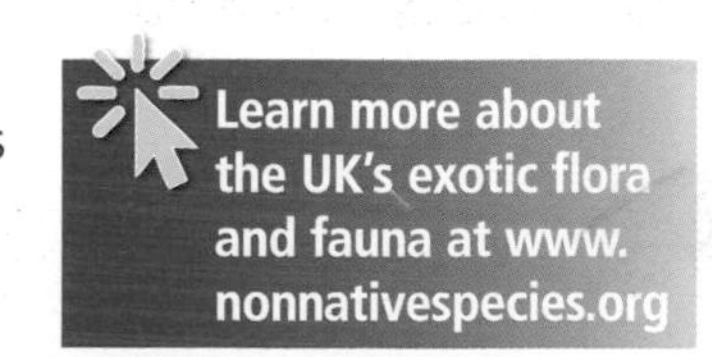

LET'S THINK ABOUT THIS

In the case of the ruddy duck, the situation is complex. It is a US native, introduced into the UK, where it has increased in population and spread to mainland Europe. In Spain, the male ruddy duck is very successful in hybridising with the closely related but highly endangered white-headed duck. To protect the European population of the white-headed duck, a cull of UK ruddy ducks is underway.

Males of the closely related ruddy duck (a) and white-headed duck (b) are both attractive mates to the endangered female white-headed ducks.

SYMBIOSIS

SYMBIOSIS MEANS LIVING TOGETHER

DON'T FORGET

Do not confuse the terms symbiosis and mutualism – not all symbioses are beneficial to both partners. Mutualism is always beneficial to both parties.

Symbiosis ('together living') refers to **intimate associations** between individual organisms of **different species**. In a symbiosis, the relationship usually provides at least one of the individuals with a **nutritional advantage**. It is not always easy to establish whether a relationship between two different species is symbiotic, but it is worth emphasising that, to be truly symbiotic, the two individuals should be living together **intimately**.

The balance of a symbiotic relationship can be affected if the host's health is reduced or environmental factors change. To reduce the harmful effects of parasites in particular, drugs and pesticides are used to promote human, animal or plant health.

PARASITISM IS A +/– SYMBIOSIS

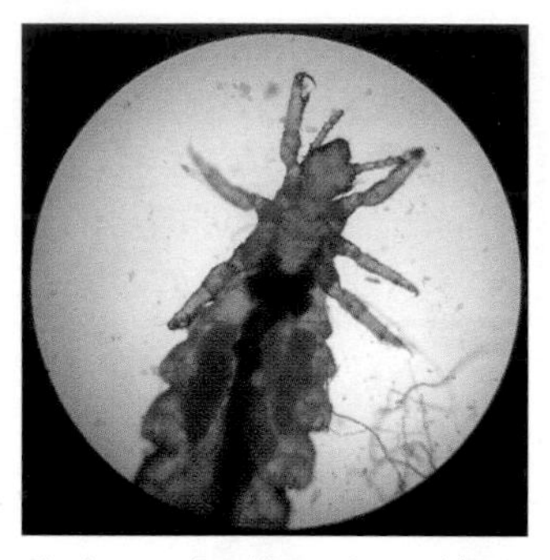

The human head louse is an obligate parasite and is highly specific to the scalp of humans. Human blood can be seen within the digestive system of the louse and its hook-like claws are perfectly adapted for clinging to fine human hair.

In parasitic symbiotic relationships, one species (the **parasite**) benefits in terms of energy or nutrients. The other species (the **host**) is harmed as it loses energy or resources. The parasite uses the host's resources for growth and reproduction; the host, as well as losing resources, also incurs further costs in defending its tissues from parasitic attack.

Adaptation of parasite and host often result in high **host–parasite specificity**. As the symbiosis is the product of **coevolution**, a relatively **stable relationship** usually exists between parasites and their hosts. After all, the fittest parasites will be those that maximise their own reproduction and transmission rather than kill their hosts; the fittest hosts will be those that can best tolerate or resist parasitic damage.

Obligate parasites are the clearest examples of symbiosis because they can only survive within the parasitic relationship. **Facultative parasites** are able to survive either within their symbiotic relationship or outside of it.

Transmission of parasites to new hosts

Parasites need to spread from host to host in order to survive. Human lice are obligate parasites and lack any resistant stage in their lifecycle – they must be passed from host to host by **direct contact**.

Human tapeworms, which live and feed in the gut, shed **resistant eggs** into the gut cavity which pass out of the host within faeces. Tapeworms return to the human host (the **primary host**) by growing and developing within the tissues of **secondary hosts**. These secondary hosts are species that (i) are likely to feed in areas contaminated by human faeces and (ii) are commonly eaten by humans (examples include cattle, pigs and fish). The risk of consuming a viable tapeworm cyst within meat is higher if hygiene is poor. In a healthy human, tapeworm infestation is often asymptomatic. However, tapeworm and other worm infections often contribute to higher death rates in developing countries.

Parasites can also be transmitted by **vectors** which move the parasite from one host to the next. One of the most feared vectors is the *Anopheles* mosquito. It transmits the protoctista *Plasmodium*, which causes malaria in humans.

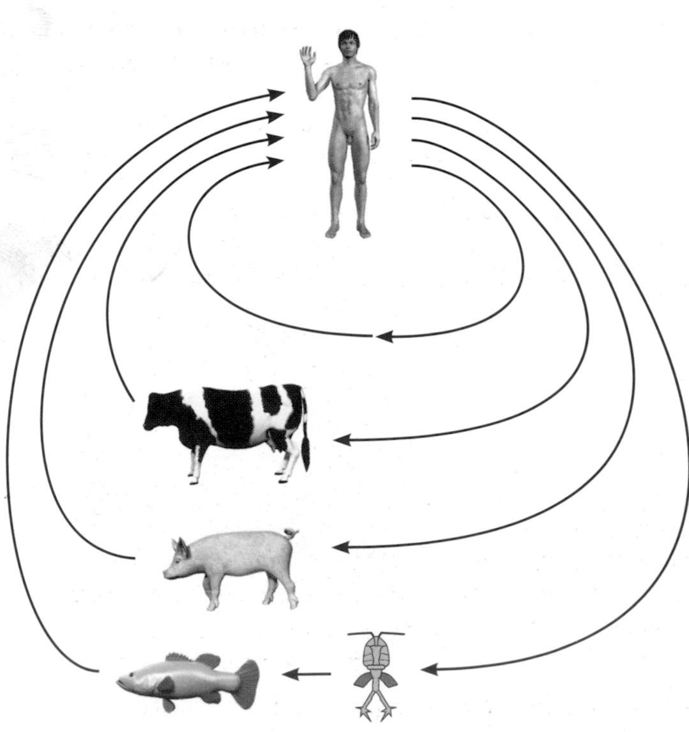

Lifecycles of four species of human tapeworm. In three species the resistant egg stages have to develop within specific secondary hosts.

COMMENSALISM IS A +/0 SYMBIOSIS

In **commensal symbiotic** relationships, one species gets a nutritional benefit (the **commensal**) and the other species (the **host**) is unaffected. A **commensalism** is usually difficult to establish with certainty, but tends to involve relationships where the commensal utilises either dead parts of the host, the host's waste or takes advantage of a disturbance in the environment caused by the host. One example of a commensal is the remora (or sharkfish) which hitchhikes on the underside of much larger fish hosts. The remora is thought to benefit as it feeds on the host's faeces; the costs to the host are thought to be negligible – the presence of the remora does not add significantly to the energetic costs of the larger fish.

Three commensal remoras and their host, the endangered shark ray; the remora can cling to the underside of a number of species of larger fish using a sucker-like adaptation on the back of its head.

MUTUALISM IS A +/+ SYMBIOTIC RELATIONSHIP

In mutualistic symbiotic relationships, **both species** involved in the interaction get a nutritional **benefit** from the relationship. As both species benefit, coevolution often generates a **structural compatibility** between the mutualistic partners, and an **exchange of metabolic products** between the mutualists often occurs (see table on the right).

Mutualistic symbiosis	Organisms and benefits
lichen	• **Algae** receive minerals from fungal breakdown of substrate. • **Fungi** receive sugars from algal photosynthesis.
coral	• **Zooxanthellae algae** receive carbon dioxide and ammonium from polyp. • **Coral polyp** receives sugars from algal photosynthesis (see page 73).
root nodules	• **Leguminous plant** receives ammonium from nitrogen-fixing *Rhizobia*. • ***Rhizobia*** receive sugars from plant photosynthesis (see page 52).
cattle stomachs	• **Protozoa and bacteria** receive fragmented food material (chewed grass) and live in ideal warm anaerobic conditions. • **Cattle** receive partially digested cellulose which can then be digested further on in their digestive system.

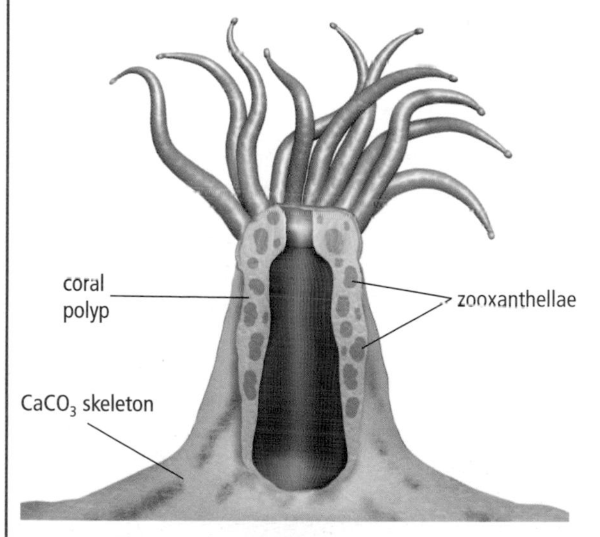

Zooxanthellae algae live within the cells of coral polyps.

Read more about symbiosis at http://www.marietta.edu/~biol/biomes/symbiosis.htm

LET'S THINK ABOUT THIS

Mutualism can also be a non-symbiotic interaction where only one organism gains nutrition, but the other gains in different ways. An example of this is the interaction between the oxpecker and large mammals. The oxpecker gains its food by eating blood-filled ticks from a **wide range** of large mammal species and these mammals benefit from reduced tick infection. The oxpecker does not have a *specific* symbiotic relationship with the white rhino species shown here, but will feed on ticks on other species too.

Oxpeckers have mutualistic but not symbiotic relationships with rhinos.

Another point to be aware of is that the dynamics of these interactions can change. For example, should ticks become scarce, the relationship ceases to be mutualistic as the oxpecker will cause harm to the host in order to feed on blood.

MANAGING INTERACTIONS BETWEEN SPECIES

UNDERSTANDING POPULATION ECOLOGY

Populations of organisms are managed for a variety of reasons. For example, the populations of both the domesticated and wild species that we use for **food or raw materials** have to be managed so that yields can be maintained at sufficient and sustainable levels. The populations of species that are of **conservation concern** are also managed, in an attempt to prevent decline and extinction.

In addition, **pest species** have to be controlled to prevent their densities reaching environmentally damaging levels. **Pesticides**, such as **herbicides** and **insecticides**, are often central to this management, but they must be used very sparingly to avoid environmental damage (see pages 70–71). Steps can be taken to manage the most important biotic interactions for each species concerned, using an understanding of population dynamics.

Population density

The management of species involves the manipulation of the abundance or, more importantly, density of species. **Population density** is the number of individuals of a species per unit space. Population density changes as a result of the effects of the **birth rate** and the **death rate**, as well as the relative rates of **emigration** and **immigration**. Obviously, a population density increases when the combined effects of the birth rate and the immigration rate are greater than the combined effects of the death rate and the emigration rate, and *vice versa*.

Learn more about population ecology at the long-named but excellent www.curriki.org/xwiki/bin/download/Coll_NROCscience/APBiologyIILesson63PopulationEcology/lesson63.zip/lesson63/lessonp.html

Biotic influences and density dependence

Density-dependent effects change birth and death **rates**. A rate can be calculated by dividing the number of births or deaths by the size of the population.

Density-dependent factors are those with an impact intensity that increases as the population density increases. That is, the effect of the factor is greater as the population density goes up. Common examples of density-dependent factors are **disease**, **competition** and **predation**. These **biotic influences** – impacts caused by living components of the ecosystem – tend to exert a regulatory or controlling effect on populations (**ecological homeostasis**). At ***A*** on the graph, the population density of the species is increasing, so the negative effects of **disease**, **competition** and **predation** will increase the death and emigration rates but decrease the birth and immigration rates; this causes a reduction in the population density. At ***B*** on the graph, the population density decreases further so the effects of predation, disease and competition are reduced, causing the population density to increase again due to decreases in death and emigration rates and increases in birth and immigration rates.

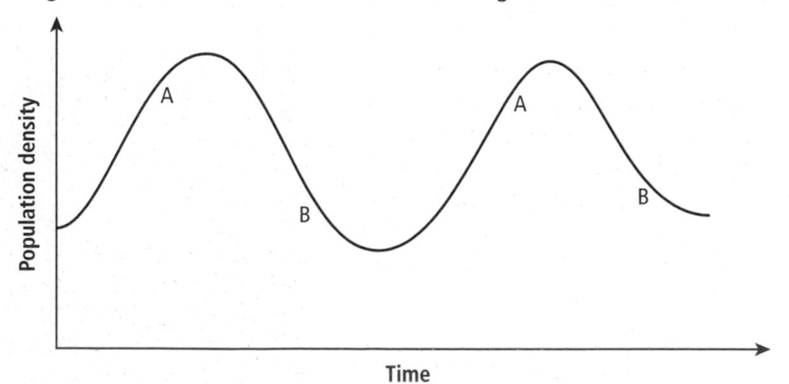

Population density varies due to the effect of density-dependent factors.

MANAGING POPULATIONS BY MANAGING INTERACTIONS

Mechanised pesticide application is a feature of intensive farming.

Managing competition

Plant competition is often managed through the use of **herbicides** to kill or control unwanted plants. To boost grain yield, selective herbicides (such as the auxin, 2,4-D) are used to destroy broad-leaved weeds. In conservation, herbicides may be used to target exotic plant species. In general, however, herbicide use tends to reduce species diversity and ecosystem stability. In Scotland, the use of herbicides is carefully regulated and is associated with intensive agriculture (see page 68).

Managing disease

Diseases in humans, other animals and plants spread more rapidly at higher population densities. Pathogens can be controlled through the use of pesticides such as **fungicides**, **antibiotics** or **antiparasitic** drugs. Other pesticides, such as **molluscicides** and **insecticides**, can be used to reduce the transmission of parasites through the destruction of secondary hosts or vectors, for example in the case of DDT control of mosquitos (see page 71). **Antibiotic** drugs are now used widely to prevent and control disease in farm animals. The liberal use of antibiotics has been shown to lead to the selection of resistant strains of bacteria.

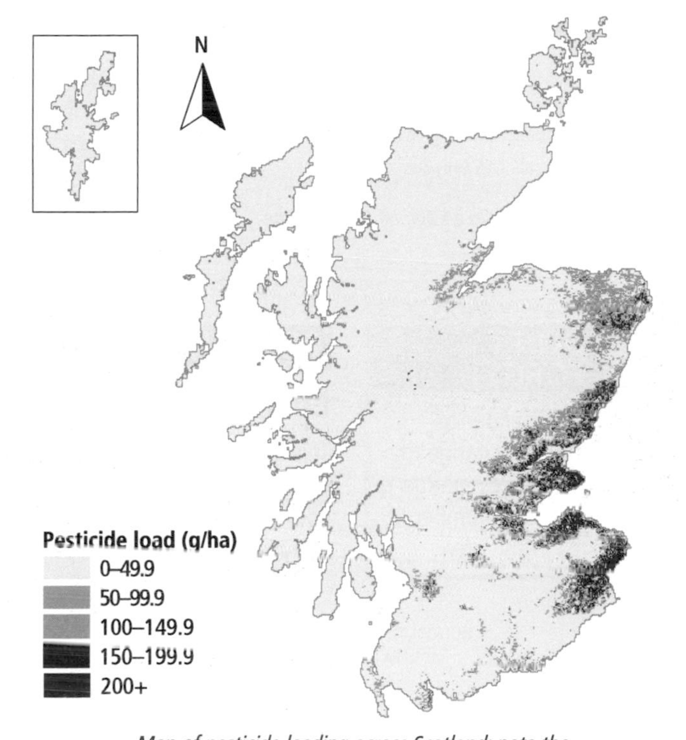

Map of pesticide loading across Scotland; note the highest use of pesticides is correlated with the areas of most intensive arable agriculture.

Managing predation and grazing

Traditionally, large predators and grazers have often been controlled in areas where crops are grown or livestock is reared. The widespread removal of these species from ecosystems has reduced diversity and caused ecosystems to become less resilient. **Pesticides**, such as **insecticides**, are often relied on to reduce species that are harmful to the crop. The substances used may be toxic and polluting (see pages 70–71) and may lead to the selection of resistant populations of insects. Agents of **biological control** can also be used – such as in the introduction of a predatory beetle to control the population of an exotic grazing insect.

Read about plans to reintroduce wolves, beavers and other species to Britain at www.islesproject.com/2009/01/31/1994-2009-wildly-ambitious-debating-the-species-to-be-reintroduced-to-britain/

LET'S THINK ABOUT THIS

Population densities of many farmland birds are now being used as indicators of environmental quality. The skylark is only one of many once-common species that have suffered serious population declines. Is it possible to *integrate* the pesticide-type approach within a complex ecosystem harbouring natural pest predators? Can we increase the sustainability of intensive farming through study of the traditional and organic farming practices that were once widespread?

The skylark, once the source of a familiar sound in the countryside, has declined greatly in the last 30 years.

CONFORMERS AND REGULATORS

INTERACTIONS WITH THE ENVIRONMENT

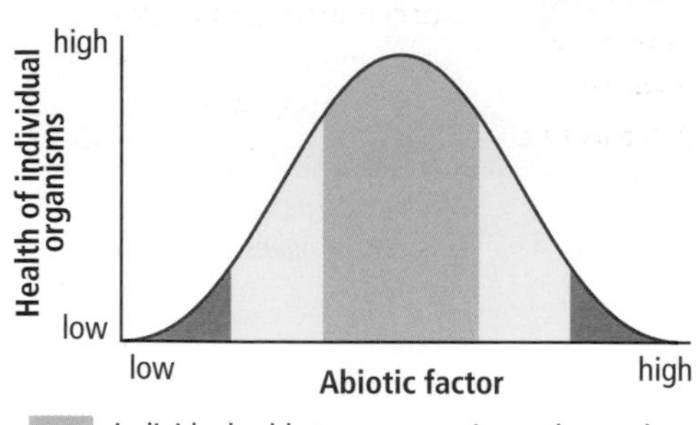

The effect of changes in an abiotic factor on the health of individuals.

Abiotic influences on organisms

Organisms are affected by **abiotic factors**. These are the non-living components of the habitat, such as water salinity or temperature. Each species is adapted to survive in a small range of abiotic conditions. When a species is not living in its optimal abiotic conditions, the population density will be lower. Abiotic factors influence organisms in two ways:

1. If the abiotic conditions in a habitat vary slightly from the optimum, this reduces the population size in the area in a **density-independent** way. This is because most of the individuals are affected in the same way, so each individual's survival chances are reduced equally.
2. If the abiotic conditions in a habitat are too far from the optimum, then the health of the individuals will be reduced. As you can see from the graph, this is a major factor in determining the distribution of a species.

DON'T FORGET

Density-independence means that the change to the size of the population is unrelated to the number of organisms in the area.

RESPONSES TO VARIATIONS IN ABIOTIC CONDITIONS

Conformation

This strategy is shown by organisms that allow their internal physiological variables to **fluctuate directly** with the changes in the external environment. **Osmoconformers** are organisms that allow their water concentration to be the same as their surroundings. **Poikilotherms**' internal body temperature is the same as their surroundings.

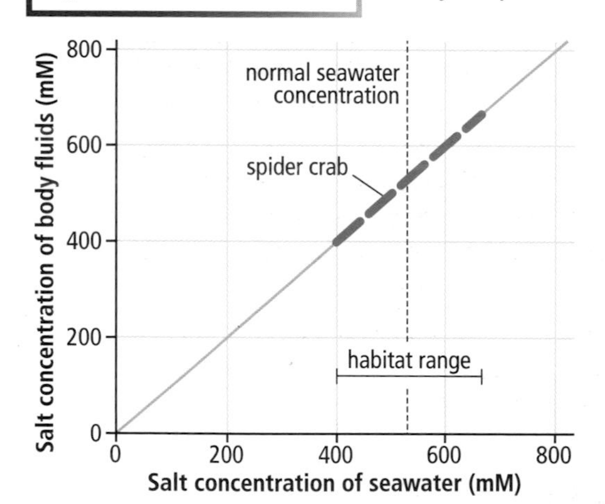

Most conformers have to stay in the habitat when the conditions change. If the change is quite small, the conformers are still able to function; this is called **tolerance**. Tolerance is only shown over a **narrow range** of the condition. If the change is too great, the organism becomes **stressed** and its survival chances are greatly reduced. In some habitats, it is possible for the conformer organism to cope using **avoidance behaviour**. For example, earthworms move deeper into the moist layers of the soil when the upper layers start to dry out.

Resistance is shown by conformers that have adaptations that isolate them from environmental change. For example, limpets resist dehydration by sealing tight to rocks.

Limited tolerance to salt concentration restricts the habitat of the spider crab.

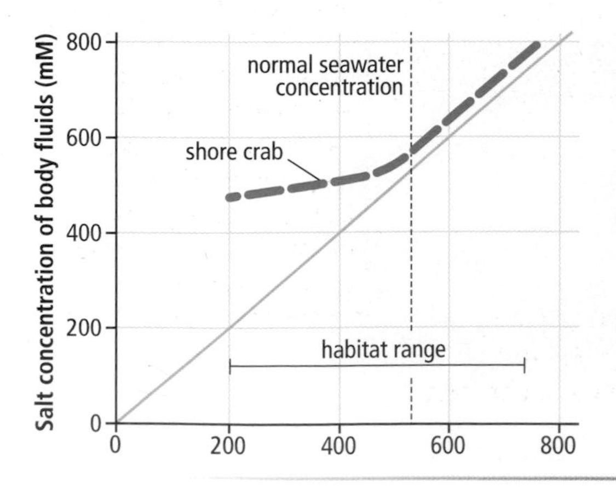

Regulation

This strategy is shown by organisms that use **homeostatic mechanisms** to maintain a **constant internal environment** despite changes in the habitat. Homeostasis is achieved through metabolic processes, so has **higher energy costs** compared with conformation.

Osmoregulators are organisms that are able to control their water balance. Freshwater fish, such as the three-spined stickleback, use **active transport** to pump salt in at their gills. Excess water is removed by the excretory system.

Osmoregulation in low salt concentrations allows the shore crab to live in a wider habitat range.

contd

RESPONSES TO VARIATIONS IN ABIOTIC CONDITIONS contd

Thermoregulators are able to keep their body temperature constant in a wide range of environmental temperatures. For this reason they are known as **homeotherms** ('same heat'). Mammals and birds are homeotherms.

Benefits of regulation

Animals that are osmoconformers are only able to occupy a **restricted range of habitats** as they are evolved to function in a narrow range of environmental conditions. As an example, starfish can only live in the marine environment. Osmoregulators can colonise a **wider range of habitats**; species of bony fish can be found in both marine and freshwater habitats.

Poikilothermic groups are also restricted in their distribution when compared with homeothermic groups. Reptiles and mammals can colonise a large range of habitats (due to their osmoregulation) but only the mammals (homeotherms) are able to cope with Arctic conditions.

Conformers have a restricted habitat occupation. Regulators are able to occupy a wider range of habitats.

DORMANCY

Some habitats are characterised by long-term changes in the abiotic conditions that are so great organisms cannot survive. The organisms that live in these habitats have evolved **dormancy**. This is a period when the organism's **metabolism is reduced**, promoting survival during the **adverse conditions**. If change occurs in a regular way (such as seasonally) organisms may show **predictive dormancy**; this occurs before the onset of adverse conditions. In habitats that are less predictable, organisms may rely on **consequential dormancy** which only starts after the onset of adverse conditions. The four main categories of dormancy are described in the table below.

Category	Examples	Description
resting spores	• microbial spores • seeds of flowering plants	The organism survives periods of drought or low temperature inside a structure that only germinates when suitable conditions return. The metabolic rate is almost zero.
diapause	• insects in temperate regions	Shorter photoperiods trigger insects to suspend their metabolism to survive the low temperatures of winter. They stay fixed in one developmental stage (for example pupa of butterflies) until conditions improve.
hibernation	• European hedgehogs • European dormice	Some small thermoregulators switch off their temperature regulation mechanism in cold conditions. This leads to a slower metabolic rate, so their body temperature falls. These organisms can survive a period of food shortage as they are not using food energy to generate heat.
aestivation	• African lungfish • desert frogs • European snails	The organism responds to the onset of very hot or dry conditions by going in to a state of torpor or inactivity. This means that its metabolic rate is greatly reduced until suitable conditions return.

Go into YouTube and type *Freezing North American Wood Frogs* to find a 3.45 minute clip about the amazing hibernating frog. This frog doesn't really fit our definition of hibernation – it's not a thermoregulator – but, as is often the way, our big definitions can't cope with the subtlety of real biology.

DON'T FORGET

Dormancy involves an active reduction in the metabolism due to genetic mechanisms. It is not merely a slowing down of enzyme activity due to colder temperatures.

LET'S THINK ABOUT THIS

Does a bear hibernate in the woods? It's true that northern bear species do sleep for long periods in winter and so conserve their energy during periods of food shortage. But their metabolic rate is kept fairly high and their core temperature only goes down by a few degrees. So, no, bears do not hibernate!

SUCCESSION

CASE STUDY: HEIMAEY, ICELAND

Go to YouTube and type *Iceland Volcano Eruption early 70s rare footage* to see more about the eruption of this volcano.

Heimaey is a small island, just off the south coast of Iceland, which has large colonies of nesting seabirds in the summer months. It is inhabited by about 4500 people.

In January 1973, a volcano erupted on the island and the town was partially covered by ash and lava. No-one was killed as there were enough fishing boats in the small harbour to ferry everyone away. The volcano erupted for about 6 months and sea water was hosed on to the lava to solidify it before it could block the harbour. After the volcano stopped, the people cleared parts of their town and re-inhabited the island.

Lava solidified just before it covered this oil storage tank at the harbour on Heimaey.

Since the eruption, **succession** has occured on the solidified lava and on the bare soil left after ash covered the vegetation. Succession is the **gradual and sequential change in the communities** present in an area over a long period of time, starting with the **pioneer community** and ending with the **climax community**. Each community replaces the previous community.

AUTOGENIC SUCCESSION ON HEIMAEY

DON'T FORGET

Facilitation is when a community alters the conditions so that they favour other species and not themselves.

The major type of succession evident on the volcanic rock and soil is **autogenic** ('self-made') as it is caused by **biological processes** of the organisms themselves. Each community alters the conditions in the habitat, so that it becomes less suited to that community and more suitable for colonisation by other species. This is called **facilitation** as the changes make it easier for a new community to colonise the habitat.

Autogenic succession follows the same **pattern of changes** in all habitats. These changes are summarised in the diagram below.

pioneer community → during an autogenic succession there are increases in:
- species diversity
- food web complexity
- niche and habitat variety

→ **climax community**

Another trend in the stages of succession is **increasing stability** – each community is more stable than the previous one. This is because the food webs become more complex through the succession, and so are more stable (see page 68). Productivity increases during a succession, but may decline as the climax community matures.

Primary succession

Primary succession occurs on areas which have not had any organisms living on them previously. First, colonisation of **barren** ground occurs. Primary succession forms soil by **producing humus**; this helps to **increase the soil depth**, the **soil nutrient content** and the **soil stability**. The succession that is still happening on the volcanic rock and ash of Heimaey is an example.

The pioneer species are **lichens and mosses** which are able to attach to the surface of bare rock or ash lumps. Lichens can form attachments using the fine threads of a fungal partner (see page 61), while mosses have root-like hairs that can get into small crevices. As pioneers, they are able to survive in very **unfavourable conditions** – they can withstand desiccation and low levels of nutrient ions.

contd

AUTOGENIC SUCCESSION ON HEIMAEY contd

The action of lichens and mosses slowly opens up any crevices. Small pockets of soil, made of dead bits of lichen and moss, start to build up in the crevices. The decaying materials **form humus** which can hold water and also contains some nutrient ions. This facilitates colonisation by more complex plants, such as leguminous plants which **increase the nitrogen content** of the simple soil (see page 52); this then facilitates colonisation by grasses. The roots of the grasses help to **stabilise the soil**. Eventually, the climax community (small shrubs and dwarf trees) may be able to colonise. This last stage has not yet happened in parts of Heimaey and may take another 50 to 100 years. The pictures below show primary succession in volcanic ash in Iceland.

Secondary succession

This occurs when the succession is on **soil that has already been formed** but which has been cleared of plant life. On Heimaey, large areas of soil were smothered by volcanic ash and all the vegetation was killed. When this ash was cleared, the exposed soil was re-colonised, initially by wind-dispersed weeds and then gradually by grasses which were able to dominate the community. In some parts of the island, the climax community of small shrubs has been able to re-colonise already. Secondary succession can happen quickly because nutrient-rich soil is already present and only needs to be stabilised by grass roots.

(a) Lichens and mosses form the pioneer community. (b) Leguminous plants colonise the small pockets of simple soil. (c) Grasses colonise the soil which has been nutrient-enriched. (d) Dwarf trees like these may be able to colonise the stabilised soil.

OTHER TYPES OF SUCCESSION

Allogenic succession ('made by others') is when the replacement of one community by another is caused by forces outside the plant community. New plants can colonise silt build-up at a river's edge; this can be considered as being an allogenic primary succession. Climate change in an area may alter some abiotic factors, meaning that some communities cannot survive there and are replaced by other communities.

Degradative succession is the sequence of detritivore organisms associated with the breakdown and decomposition of a dead animal or plant. It happens in a **much shorter time frame** than the other types of succession. Each detritivore community uses different food sources in the organism, and later communities can only feed once other parts have been removed. This process is also known as a **heterotrophic succession** because it only involves organisms that gain complex organic molecules by consuming them – there are no plants in the succession. For example, dead sea birds on Heimaey will be fed on by fly maggots within a few weeks and it may take three years before hide beetles move in to feed on the skin from the inside, finishing the succession.

A good interactive explanation of primary and secondary succession can be found at http://ilo.ecb.org/SourceFiles/succession.swf

DON'T FORGET

Make sure that you are clear about the differences between autogenic and allogenic succession.

LET'S THINK ABOUT THIS

Degradative succession is very different from the plant-based autogenic successions. Species diversity, food web complexity and niche diversity do increase through the first part of the succession, but then gradually decrease to nil.

The food supply is gradually being used up by each community so there is less diversity of food available until eventually there is no food left to support any detritivores.

INTENSIVE FOOD PRODUCTION

THE REQUIREMENT FOR INTENSIVE FOOD PRODUCTION

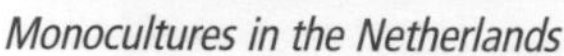

Monocultures in the Netherlands

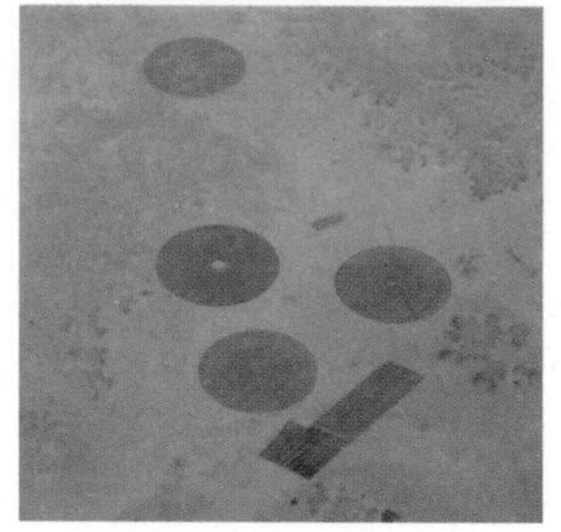

Intensive farming in the Sahara Desert – a mechanised rotary irrigation system in Libya

Global food production has to keep pace with **human population increase** (the human population will exceed 7 billion by 2011 or 2012). To meet this demand, farming has had to become **intensive**. Intensive farming uses **high inputs per unit area** to **maximise yield**.

Achieving high yields

Typically, intensive food production focuses on growing **high-yielding varieties**. In the cultivation of crops such as rice, wheat and maize, for example, varieties can be the products of careful artificial selection, including transgenic modifications (see pages 44–45).

To increase yield further, **optimal nutrients for rapid growth are provided**, such as in the processed foodstuff given to pigs, cattle and chickens. Other growth promoters, such as bovine somatotrophin (BST), may also be used (see page 44). Intensive cultivation of plant crops uses synthetically produced macronutrients such as inorganic ammonium fertiliser manufactured using the Haber process.

Greater efficiency of intensive farming can often be achieved through the cultivation of crops as **monocultures** – single varieties or single species grown together on a large scale. The uniformity of the crop in monocultures is often increased through the use of F_1 hybrids or micropropagation by tissue culture (see page 11). Habitats are modified, such as in the removal of hedges to create larger fields or the conversion of mangrove habitat to tropical aquaculture.

Advances in **pest and competitor control** allow for regular applications of pesticides such as insecticides, fungicides, molluscides, herbicides and antibiotics. These ensure that yield is high by boosting crop productivity and limiting productivity of non-crop species.

Intensive farming relies on fossil fuel-powered **mechanisation** of soil and crop management. Large-scale monoculture requires the simultaneous mechanical harvesting of large areas of crop.

IMPACTS OF INTENSIVE FARMING ON THE ENVIRONMENT

Loss of species diversity and ecosystem complexity

Intensive food production **reduces species diversity**. High crop productivity is achieved at the expense of native species – biodiversity that interferes with the crop tends to be eliminated. Large-scale monoculture creates a uniform habitat, and pest and competitor control restricts the non-crop ecosystem. This reduced species diversity results in a **loss of ecosystem** complexity – there are fewer species, so there are fewer niches and fewer links within the food web. A restricted ecosystem has **lower stability** than most natural ecosystems and this can lead to problems. For example, the genetic uniformity of the crop risks the rapid proliferation of any pest that does gain a foothold, and the lack of niches in the ecosystem makes it unlikely that a natural predator of this pest would exist within the crop.

Compare a complex natural habitat with one produced by intensive farming. (a) A complex mosaic of natural habitat in the South African savannah can support complex food webs, including megafauna (such as the martial eagle and white rhino shown here). (b) Intensive cultivation of olives for oil in Europe involves the mechanical and chemical suppression of competition between trees, creating a simple and potentially unstable ecosystem.

contd

IMPACTS OF INTENSIVE FARMING ON THE ENVIRONMENT contd

Poorer soil condition

Intensive farming often damages one of the most precious natural resources for growing food – the soil. Single-species cropping combined with mechanisation results in a **poorer soil condition**. Large machinery causes compaction of the soil crumb structure, which can increase erosion and soil loss. Repeated cropping of the same species from the same area exhausts the supply of certain nutrients in the soil and may also reduce the organic content of the soil.

Algal bloom in an upland lochan caused by eutrophication through run-off contaminated with animal waste.

Increased use of artificial fertilisers

The addition of chemical fertilisers containing inorganic nitrates and phosphates is common practice in intensive food production. If the application of fertiliser is not managed correctly, rainfall can lead to the loss of these nutrients either by **leaching** (when dissolved nutrients wash through the soil) or as **run-off** (nutrients become dissolved in surface water and are washed away). When inorganic nutrients are washed into freshwater ecosystems **eutrophication** may occur, which can lead to environmental damage by **algal bloom**.

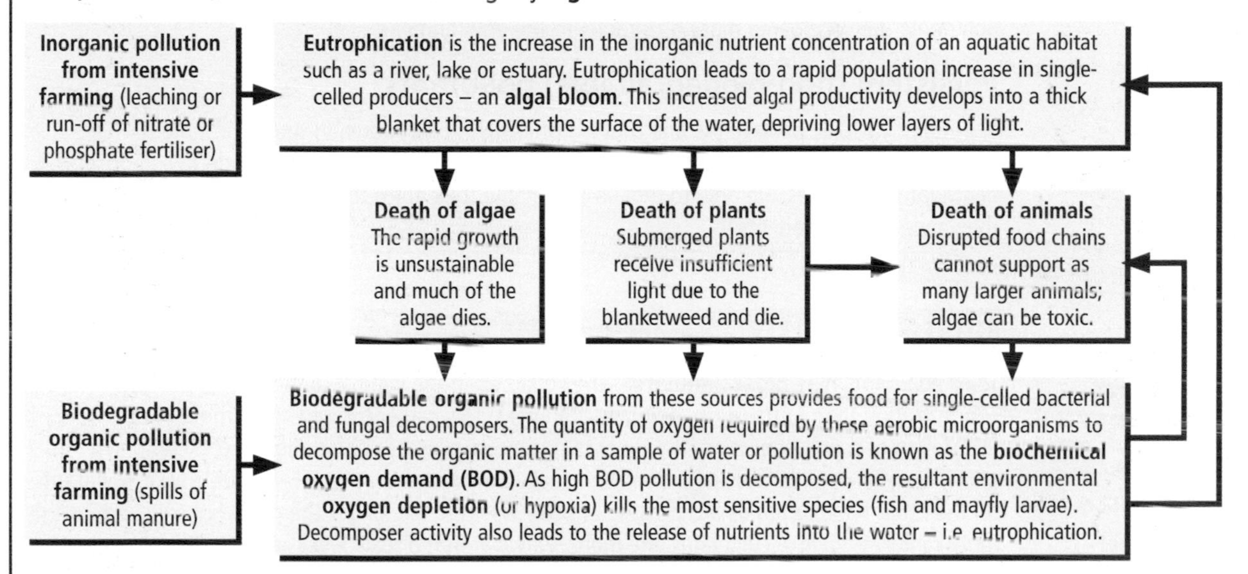

The development of oxygen depletion as a result of inorganic or organic pollution

DON'T FORGET

Eutrophication is *not* another word for an algal bloom, but an increase in nutrient concentration that can *lead to* algal bloom.

Biodegradable organic pollution and BOD

Biodegradable organic pollution (such as animal manure or plant waste) can cause great damage to freshwater ecosystems. Any pollutant with a high **biochemical oxygen demand (BOD)** will stimulate the rapid growth of aquatic decomposer bacteria and lead to a low concentration of dissolved oxygen (as shown in the diagram). This oxygen depletion causes the death of some aquatic animals. More **susceptible species** (such as mayfly larvae) die out rapidly in these circumstances. Other, more **favoured species** (such as sludgeworms) may thrive with a lack of competition. The relative abundances of these **indicator species** (the sensitive and resistant species) can be used to monitor the quality of the environment.

Increased use of pesticides

Intensive farming is associated with an increased use of pesticides. By their nature and intention, pesticides are toxic to components of the ecosystem and their inappropriate use can have serious environmental consequences. For more details of the **bioaccumulation** and **biological magnification** of **persistent** pesticides see pages 70–71.

DON'T FORGET

Many potential pollutants from intensive farming, such as silage run-off or pig slurry, contain a wide mixture of both inorganic and organic pollution.

LET'S THINK ABOUT THIS

Intensive farming has been successful in raising food production to a level that can provide enough food for all. However, poverty and conflict mean that an estimated **13% of the world's population is undernourished** in terms of energy and/or protein. This undernourishment is the cause of more human deaths than any other single factor.

BIOACCUMULATION

TOXIC POLLUTION

Human activity that releases unwanted and damaging substances into the environment is known as pollution. **Toxic pollution** refers to substances that are harmful to the health of organisms.

Toxicity and specificity

The higher the **toxicity** of a chemical, the more damage it can cause if discharged into the environment. Toxicity increases with dose concentration and length of exposure, and can be influenced by the age, health and genetic predisposition of the organism concerned. Some pollutants, such as heavy metals, are toxic to a wide variety of organisms. Others are only highly toxic to **specific** (sensitive) groups of organisms. For example, the spillage of a herbicide tends to cause much higher levels of direct damage to plants than animals. In general, **non-specific** pollutants are considered more hazardous to the environment.

Persistent (non-biodegradable) pollutants

A **persistent** toxic pollutant is one that is not easily broken down in the environment. These **non-biodegradable** molecules, being resistant to the actions of consumer and decomposer degradative enzymes, are able to remain in a stable toxic form for long periods of time (that is they have a **long half-life**).

BIOACCUMULATION

DON'T FORGET

Bioaccumulation refers to the higher concentration of pollutants within an organism, trophic level or community compared to the concentration in the environment.

DON'T FORGET

Biological magnification refers to the accumulation of pollutants to higher and higher concentrations in successive trophic levels within an ecosystem.

Toxic pollutants may **bioaccumulate** – they build up in the cells of organisms to a higher concentration than in the surrounding environment. This can occur whenever the rate of absorption of a pollutant is greater than its combined rates of degradation and excretion. In animals, water-soluble toxic pollution can usually be excreted in urine. Fat-soluble toxic pollutants, on the other hand, often bioaccumulate in **fatty tissues**. The longer the period of exposure or the more rapidly a toxin accumulates, the more likely it is that a toxic level will be reached.

Biological magnification

Toxic pollutants that bioaccumulate can sometimes be found in unexpectedly high concentrations in successive trophic levels of an ecosystem. This is called **biological magnification** and it is caused by the accumulation of the toxic pollutant at each trophic transfer. The small biomass present at higher levels in the ecosystem is supported by the consumption of a large biomass at lower levels, and this transfer increases the concentration of the toxic pollutant at each level in a food chain. The consequence of this biological amplification is that low or harmless initial environmental concentrations of a pollutant rapidly become toxic or **lethal in top predators**. Examples of toxic pollutants that show biological magnification are the **insecticide DDT** and **mercury** (a heavy metal).

Biotransformation

Once a toxic pollutant has entered an organism, there is a chance that it may undergo various chemical changes. This **biotransformation** may degrade the pollutant successfully, or may result in the formation of another toxic chemical. Examples of the latter include the biotransformation of DDT to DDE in birds and the biotransformation of elemental mercury (Hg) to biologically more toxic methyl mercury (CH_3Hg) by anaerobic bacteria.

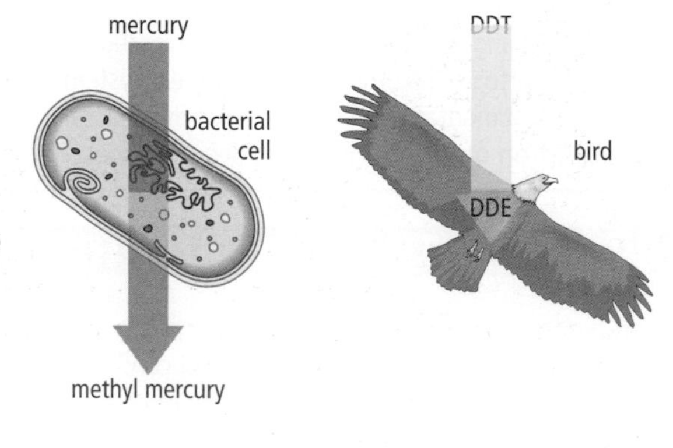

CASE STUDIES OF TOXIC POLLUTION

Mercury

Mercury, like lead, cadmium and chromium, is a **heavy metal**. Mercury is a **persistent** (non-biodegradable) pollutant, so has become widely distributed in the environment as a result of industrial discharge. Mercury is an irreversible non-competitive inhibitor of enzymes (see page 27). Like other heavy metals, mercury **bioaccumulates** and it is known to affect nervous tissue and foetal development.

An infamous example of long-term pollution of the environment with mercury occurred at Minamata in Japan between the 1930s and 1960s. Discharge of mercury led to **biological magnification** through the marine food chain. Thousands of people were affected and the horrific symptoms of methyl mercury poisoning in humans are now known as **Minamata disease**.

Concentration of methyl mercury (parts per billion)					
water	detritus	detritivore	minnow	sunfish	bass
0·0005	0·1–0·5	0·2–0·8	2–200	500–1000	500–5000

Biological magnification of mercury in a detritus-dependent aquatic food chain

How widespread is mercury pollution? Use the website www.newscientist.com and search for *Faroe islanders told to stop eating 'toxic' whales.*

DDT

DDT (dichlorodiphenyltrichloroethane) is a potent **non-specific insecticide** – it interferes with nerve impulse transmission in insects. It is a **persistent** pollutant with a **long half-life** as it is almost **non-biodegradable**. It is found in all ecosystems, even though its use has been restricted for many years. It **bioaccumulates**, especially in **fatty tissues**, and **biotransformation** into DDE also occurs. Both DDT and DDE are harmful to birds at high concentration and, due to **biological magnification**, the toxic effects tend to be seen in top-level predators. DDT and DDE reduce calcium deposition during egg formation. The thin-shelled eggs which result are more easily broken in the nest and the lower hatching and survival rate of chicks has resulted in regional extinctions of birds such as the osprey.

DON'T FORGET

Make sure you are clear about the differences between the terms bioaccumulation, biotransformation and biological magnification!

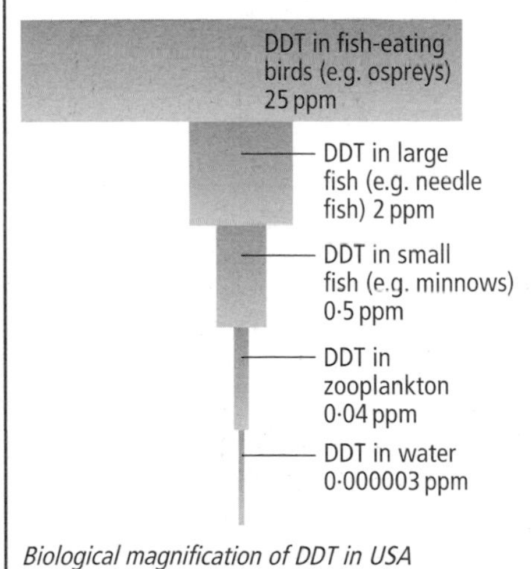

Biological magnification of DDT in USA

Osprey

Find out how the environment will be protected from the effects of DDT biological magnification by reading *10 things you need to know about DDT use under The Stockholm Convention* on the WHO website at www.who.int/malaria/ddtandmalariavectorcontrol.html

LET'S THINK ABOUT THIS

The discovery of the damaging effects of DDT on bird populations was highlighted in the early 1960s by Rachel Carson in a book called *Silent Spring*. This had a major influence in starting the Green Movement in the USA and eventually led to most countries banning DDT use in the 1970s and 1980s. However, the ban had to be reversed by the World Health Organization (WHO) in 2005 as DDT had been essential in combating deaths from malaria in developing countries and the increase in prevalence of malaria was becoming a major health issue.

AIR POLLUTION

FOSSIL FUELS

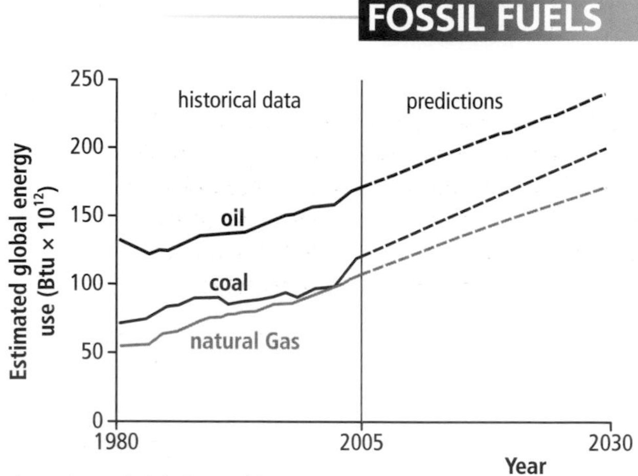

The estimated global use of fossil fuels. ('Btu' is the industry unit used to measure the heat released by a fuel.)

Fossil fuels are a finite resource

Millions of years ago in some parts of the world, huge masses of dead organisms were trapped during rock formation. The pressure exerted by the rock converted these remains into fossil fuels: **coal, oil and natural gas**. The laying down of fossil fuels happened for only a short period of geological time, so these are **finite resources** that will run out.

Since the 19th Century, the **burning of fossil fuels** has been the major source of energy for the maintainence of lifestyle and agriculture in the industrialised world. The **demand for energy is growing** as people in the developing nations (such as India, Brazil and China) become more affluent and seek to improve their standard of living (see graph).

There is clearly an urgent need to **conserve fossil fuels**. This will require the development of more **energy efficient** technologies and the use of **alternative sources of energy** (wind, tidal, nuclear, hydro and solar).

Air pollution from fossil fuels

Burning fossil fuels increases air pollution by causing the **emission of gases** into the air. These gases can be classified in two groups.

DON'T FORGET

Burning fossil fuels releases three acidic gases and two greenhouse gases. Carbon dioxide fits in both categories.

1. **Acidic gases** (sulphur dioxide, nitrogen oxides, carbon dioxide) – these **dissolve in atmospheric water** to form **acid rain**. When it falls, acid rain lowers the pH of soil and water ecosystems; changes to abiotic conditions can affect the health and survival of some species (see page 64). Acid rain also dissolves many of the nutrient ions out of the soil.
2. **Greenhouse gases** (carbon dioxide and water) – these enhance the greenhouse effect leading to global warming (see below).

GLOBAL WARMING

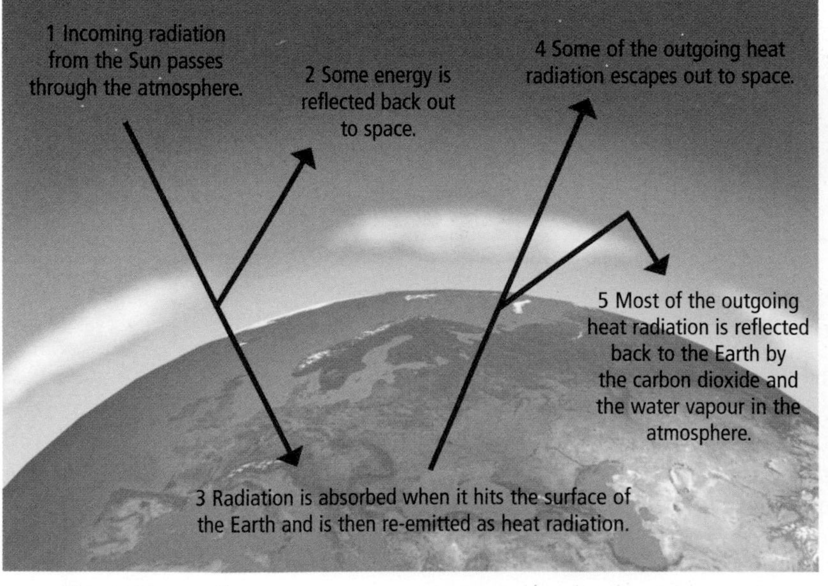

The greenhouse effect

The greenhouse effect is normal

The greenhouse effect is an essential part of the ecology of the Earth. It helps to **maintain the temperature** of the atmosphere at a level that is suitable for life on Earth. Estimates suggest that without the greenhouse effect the average global temperature would drop from 14°C to –18°C!

The greenhouse effect is caused by **normal carbon dioxide and water vapour** in the air. Most of the **incoming radiation** from the Sun passes through the atmosphere. This radiation is absorbed when it hits the surface of the Earth. The energy is re-emitted as heat, a little of which escapes out to space. However, most of this **outgoing radiation is reflected** back to the Earth by the carbon dioxide and the water vapour.

contd

GLOBAL WARMING contd

The enhanced greenhouse effect is caused by us

If the level of carbon dioxide in the atmosphere increases due to human activity, more heat is reflected. This is called the **enhanced greenhouse effect** and it is mainly due to the **increase in carbon dioxide** from burning of fossil fuels. Other pollutants can also enhance the greenhouse effect:

- **Chlorofluorocarbons (CFCs)** were used in refrigerators and freezers and can be released into the atmosphere if these machines are not properly decommissioned. One CFC molecule has the same warming effect as 10 000 molecules of carbon dioxide.
- **Methane** is released by anaerobic microorganisms; some of these live mutualistically in the stomachs of cattle (see page 61) while others are free-living in water-logged soils, such as rice fields. This pollutant is increasing by about 2% per year due to increased agriculture. Although each molecule is about 20 times as potent as carbon dioxide, it is not as long-lived in the atmosphere.

The enhancement of the greenhouse effect is increasing as human activity produces more air pollution. This is causing the global temperature to increase, an effect known as **global warming**. It is believed to be responsible for **dramatic changes to the climate,** with some parts of the world having more hurricanes and floods, while others have had unprecedented periods of drought.

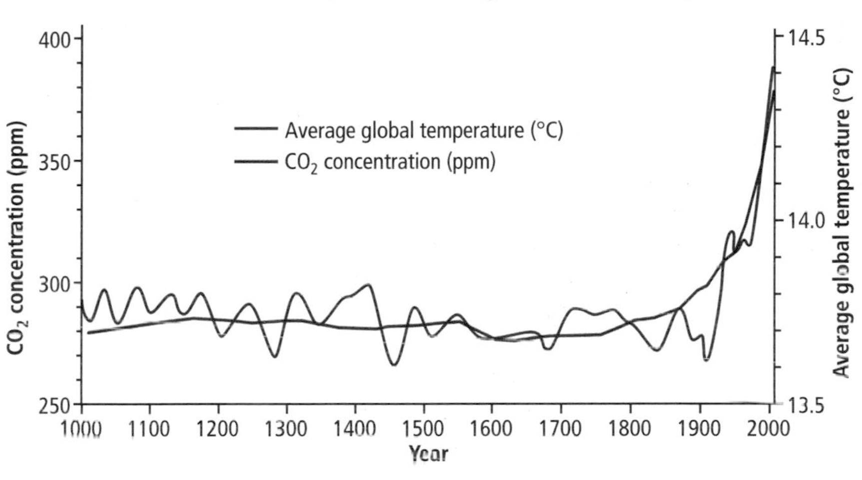

The link between CO_2 concentration and global warming

GLOBAL WARMING AFFECTS ORGANISMS

It is estimated that the average global temperature will increase by 2°C to 5°C by the end of the century. This will cause **local habitat destruction** and, with each 1°C rise being equivalent to a change of 100 km to 150 km of latitude, it is clear that global warming will **affect the distribution** of many species.

Examples of affected organisms

- **Coral bleaching** – Coral polyps form a mutualistic relationship with a type of alga called **zooxanthellae** (see page 61). The algae gain carbon dioxide and ammonium from the coral polyp, and the polyp gains sugars from algal photosynthesis. A sea temperature increase of even 1°C can upset the balance of this relationship, causing the build up of harmful algal metabolites. The polyp expels its now toxic partner. As a result, the coral polyp dies leaving only its limestone skeleton, hence the term '**coral bleaching**'.
- **North Sea fish** – from 1977 to 2001, the average temperature of the North Sea rose by 1°C. Two-thirds of fish species moved their ranges further north to remain in a favourable habitat.

Coral bleaching is the key example to remember for the exam.

A great site explaining more about coral bleaching, with clear text and a film is at http://www.stanford.edu/group/microdocs/coralbleaching.html

LET'S THINK ABOUT THIS

Global warming is melting the polar ice but there could be a bigger problem due to the melting of the permafrost. This frozen peat bog covers most of the far northern latitudes and is full of methane formed by anaerobic bacteria. When the permafrost thaws, there will be a large release of methane, causing a 'tipping point' where global warming will rapidly increase.

USE OF MICROORGANISMS

Biotechnology is the use of living organisms to make **products of value to humans**. Traditionally, biotechnology has been used in the production of **food** (such as in brewing and baking). More recently there have been great developments in the use of biotechnology in **medicine**, **agriculture**, **tissue culture** and the industrial production of molecules such as **enzymes**.

Microorganisms, such as bacteria, single-celled fungi and protoctistan species, are often used for biotechnology. Their relatively simple culture requirements mean that they can often be grown in large numbers in pure cultures and their products or the organisms themselves can then be harvested.

GROWING MICROBES

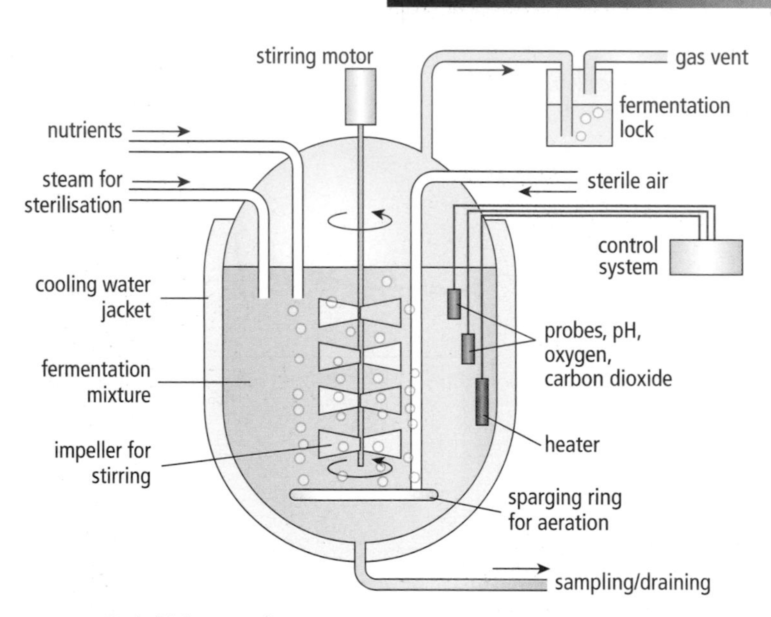

Typical laboratory fermenter

Microorganisms are grown in a **culture medium**. The medium contains all the nutrients required by the microorganism, such as inorganic salts, and, depending on the species, may include organic sources of energy, carbon and nitrogen. Cultures of fungal cells require a **complex medium** which includes ingredients such as malt extract.

Culture medium can be mixed with agar to make a solid or semi-solid surface on which to grow colonies of microbes, such as in a Petri dish. Alternatively, microorganisms can be cultured within the liquid medium in a fermenter vessel. Either way, other aspects of the growth conditions must be controlled carefully, such as gas concentrations, pH and temperature. **Fermenter vessels** usually allow the culture conditions to be monitored and controlled. In addition, once product formation has occurred, outlet ports allow the removal of medium for product purification and recovery.

Aseptic technique

Aseptic technique is a set of precautions taken to prevent contamination. It is important to avoid contamination of the culture vessel by unwanted species and to also avoid contamination of the laboratory or personnel by the organisms being cultured. Aseptic technique reduces the risk of culturing hazardous species unintentionally.

Key to aseptic technique is the use of impervious vessels such as plastic Petri dishes or glass or stainless steel fermenters for the **containment** of the organisms. These vessels can be **sterilised** by radiation or by a combination of heat and pressure to ensure that no contaminating organisms (for example, viable bacterial or fungal spores) are present. Contamination risk is also reduced by the use of personal protective equipment, such as a lab coat and gloves, and by **disinfection** of the working area.

Microbiological techniques are usually performed within the updraft of air created by a lit Bunsen burner, which reduces the chances of airborne contaminants falling on the work space. Metal and glass implements are sterilised either by red heat in the Bunsen flame or by dipping in ethanol and burning off the alcohol. The necks of all tubes and bottles are passed through the Bunsen flame before and after all operations.

Fermentation vessels, such as these ones in a whisky distillery, must be sterilised using steam between batches.

DON'T FORGET

Make sure that you know the difference between **disinfection** and **sterilisation**.

contd

GROWING MICROBES contd

Both fungi and bacteria produce dormant spores to combat environmental adversity (see page 65). Many of these spores are resistant to damage, even from boiling water. To be sure of destroying these spores during sterilisation procedures, a dry heat of around 150°C is required for around 2 hours. Alternatively, an autoclave can be used to raise steam temperature to 121°C under pressure for 15–20 minutes.

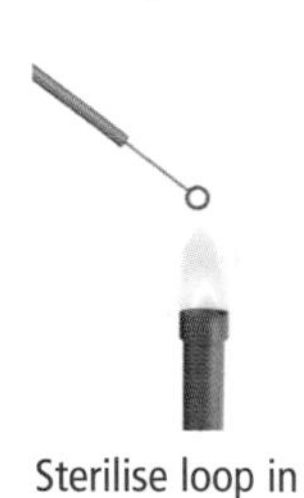

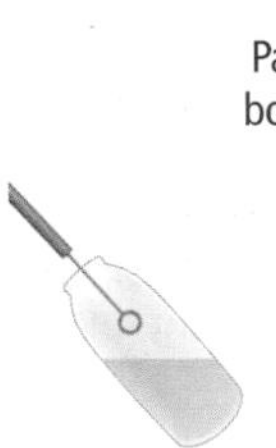

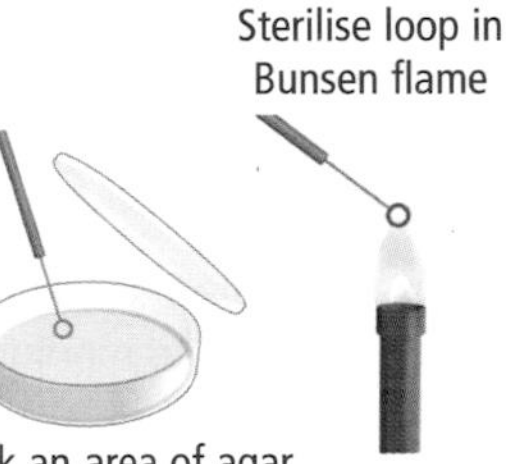

Stages of subculturing a microorganism from a pure colony on an agar slope, using aseptic technique

Obtaining pure cultures

As an essential precaution to avoid contamination and ensure the purity of microbial cultures, methods have been developed to isolate individual colonies. The commonest technique is the use of a streak plate. In **streak plating** an implement, such as a metal loop, is repeatedly sterilised and used to drag a sample across successive areas of an agar plate. With each successive streak, fewer and fewer microorganisms are moved into each new area. Once incubated, pure individual colonies are usually found growing along some of the later streaks. These pure cultures can then be used as a source of **inoculum**.

Once a pure culture has been isolated in this way and used to inoculate fresh medium, a microbiologist could use the same streak plate technique to check the purity of any sub-culture – any contaminating species would tend to show up on the streak plate as colonies with a different growth form.

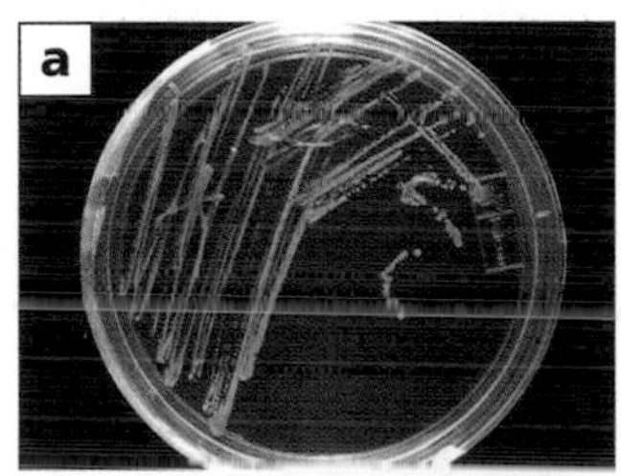

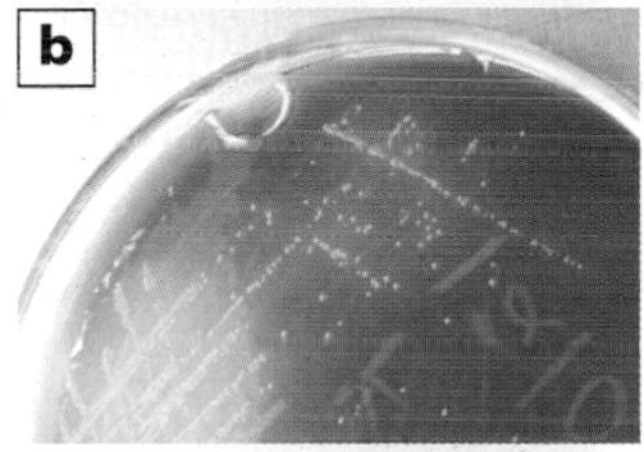

Single colonies are isolated using streak plates: (a) Micrococcus luteus *on nutrient agar; (b)* Vibrio cholera *on bile salt agar; (c)* Escherichia coli *on MacConkey agar.*

Scaling up

Several factors have to be taken into account when scaling up culture from Petri dish to full-scale industrial production. Normally, a **laboratory fermenter** has a culture volume of about 0.2 litres. A **small-scale pilot fermenter** of about 200 litres would be used to establish the various optimum conditions (such as temperature and pH) as these might differ from the laboratory scale. Only once these conditions have been determined can **full-scale industrial production** be attempted. Industrial production may use vessels that are many thousands of litres in capacity. The control of the temperature of these vessels, for example, is much more difficult than at smaller scale. It is essential that the scaling up is carried out carefully, if expensive mistakes at the industrial scale are to be avoided.

Industrial-scale fermenters of several thousand litres' capacity as used in wine production

Small-scale pilot fermenter of about 100 litres' capacity.

LET'S THINK ABOUT THIS

Two considerations when culturing microbes in the laboratory are that:

- temperatures close to human body temperature (37°C) are avoided
- anaerobic culture is used sparingly.

This is to avoid conditions that may encourage the growth of pathogens.

GROWTH OF MICROORGANISMS

MEASURING MICROBIAL GROWTH

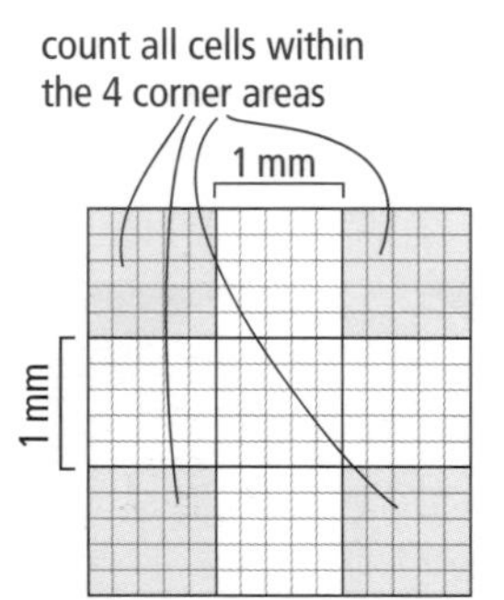

A haemocytometer slide is used to make direct cell counts under the microscope.

A population of microorganisms can be **measured directly** using cell **counts**. High-powered microscopy using haemocytometer slides allows accurate cell counts in a standard volume of culture. Staining methods can be used to differentiate between **viable** (live) and **total** (live and dead) cell numbers.

Alternatively, viable cell counts can be made by **plating** known volumes of a series of **dilutions** of a culture onto solid medium and then counting the number of colonies that grow. Using serial dilutions, it is likely that some plates will be countable.

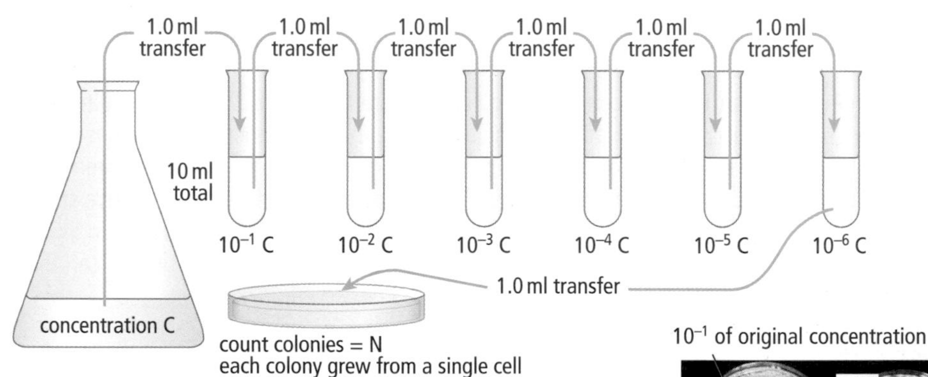

A dilution series involving six 10-fold dilutions. Samples from each dilution are plated to make viable counts. In this example, plates at 10^{-4} and 10^{-5} dilutions have countable numbers (N) of colonies. From these two dilution counts, the undiluted concentration of microbes (C) in the original culture can be calculated to be approximately 2.6×10^7 cells per ml.

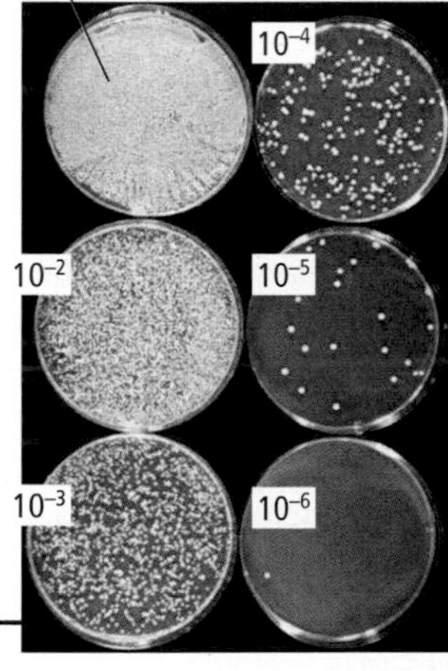

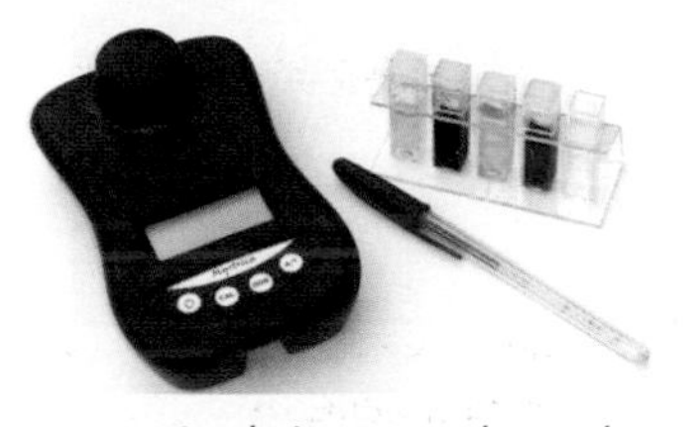

A colorimeter can be used to estimate total cell densities.

Cell populations can be **estimated indirectly** using **turbidometric** methods. The increase in absorbance (or reduction in transmission) as measured by a colorimeter can be used to estimate the increasing density of cells in a culture. Clearly, this method can only estimate total cell counts. Both live and dead cells contribute to turbidity.

STAGES OF GROWTH

Lag phase

After inoculation, growth is **slow** initially. While cells may increase in size, there is little increase in cell numbers. The low rate of cell division can be the result of the cells in the inoculate not being in the actively dividing log phase. Alternatively, if the culture medium is different from the inoculate medium, the microorganisms may need time to **synthesise enzymes** before population growth can begin.

Bacterial growth is measured as population growth and the rate of increase can be so great that it usually is recorded on a logarithmic scale.

Log phase

The log phase is also called the **exponential** phase and it is during this period that the most rapid population growth of the microorganism occurs. The **maximum rate** of reproduction reached during this phase depends on the supply of nutrients and how close the conditions are to the optima. The term exponential refers to a doubling of cell numbers with each generation. The time to produce a new generation, the **generation time** or **doubling time**, is represented by **g**.

Once the doubling time (g) has been established from cell counts, the **growth rate constant** (**k**) can be calculated. The growth rate constant is a measure of the number of doublings that occur in unit time.

contd

STAGES OF GROWTH contd

Since population growth in the log phase is exponential, the growth rate constant can be calculated using the following equation:

$$\mathbf{k} = \frac{\text{natural log2}}{\text{doubling time}} = \frac{\ln 2}{g} = \frac{0.693}{g}$$

By calculating **k**, different cultures can be compared and growth optimised.

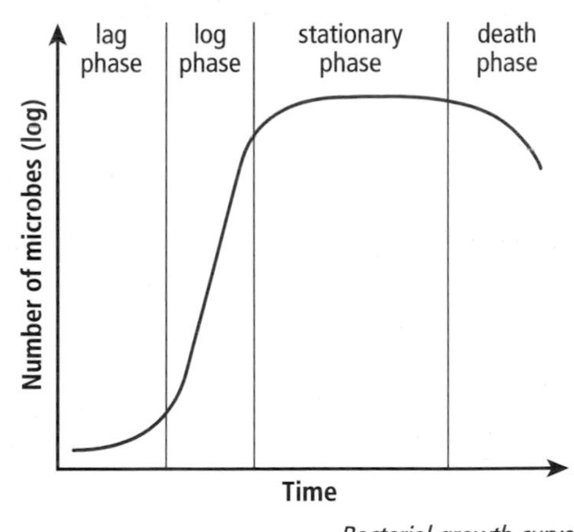

Bacterial growth curve

Stationary phase

Population **growth halts** when the death rate of cells is in balance with the production of new cells. This will occur when growth is limited by either low nutrient availability or the build-up of products of metabolism in the culture medium. It is during the stationary phase that many important secondary metabolite molecules are produced (for penicillin, see page 84).

Death phase

Population decline occurs once the rate of cell death outstrips the rate of new cell production. The process of **autolysis** (the self digestion of cells by lysosome action) can release nutrients for some further cell growth and can also provide useful products for harvest by biotechnologists.

THE *LAC* OPERON AND DIAUXIC GROWTH

An unusual bacterial growth curve is the diauxic growth curve. This **two-phased growth** curve occurs when bacteria are given two food sources, one of which is utilised before the other, such as *E. coli* feeding on a mixture of glucose and lactose. Only once the glucose is metabolised are the genes for lactose metabolism switched on.

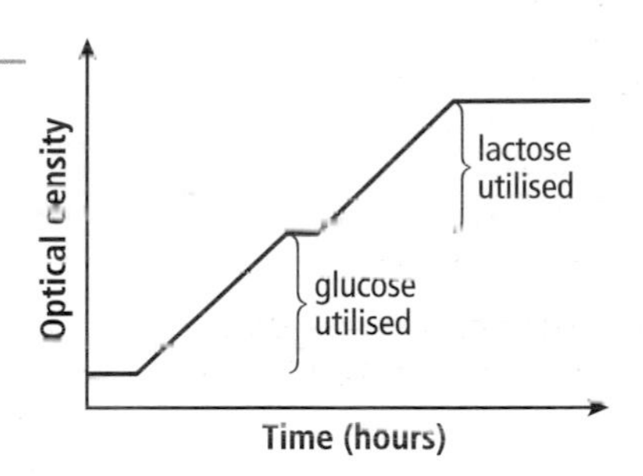

Diauxic growth curve

The switching on and off of the gene for lactose digestion is controlled by the binding of repressor protein at the **operator sequence** and activator protein at the **promoter sequence**. The presence of lactose acts as an **inducer** and can stimulate β-galactosidase production, but only in the absence of glucose (**catabolite repression**).

A regulator gene synthesises a repressor protein, which binds to the operator, twisting the DNA and preventing transcription. The repressor also has a binding site for lactose, and when lactose is present the repressor is unable to bind to the operator. However, this is not enough to switch on transcription. A protein known as CAP must first bind at the promoter sequence of the gene. CAP will only bind in the presence of cAMP, and cAMP only builds up in the cell in the absence of glucose. So, the gene for lactose digestion is not transcribed until glucose levels are low.

An operon consists of a structural gene, its promoter and its operator.

a

With glucose, little cAMP
Without cAMP, CAP does not activate

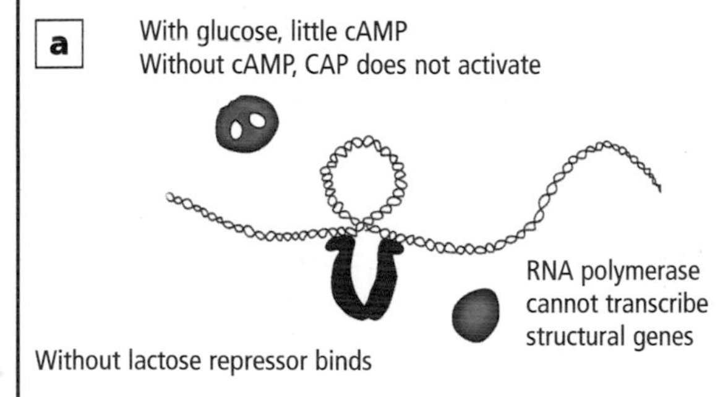

b

Without glucose, cAMP increases
With cAMP, CAP binds and activates

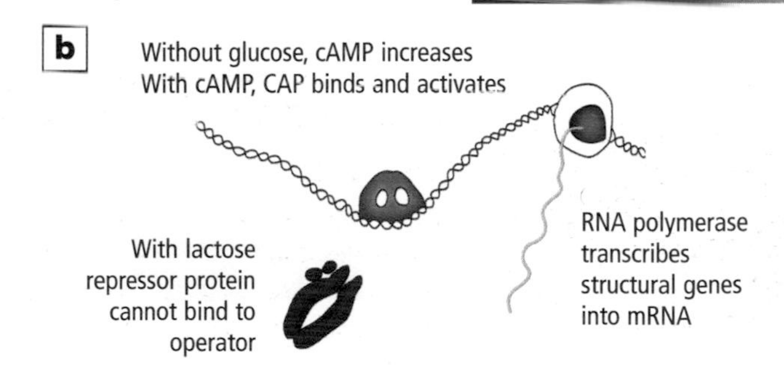

(a) Lac *operon in presence of glucose and absence of lactose: glucose metabolism results in low cAMP levels in the cell; cAMP does not bind with CAP (orange) and CAP fails to bind to the promoter. Without lactose, the repressor (red) binds to the operator. No transcription.*

(b) Lac *operon in the absence of glucose and presence of lactose: high cAMP levels allow CAP (orange) to bind at promoter sequence. With lactose present, repressor protein (red) cannot bind to the operator. This combination results in successful transcription.*

LET'S THINK ABOUT THIS

The *lac* operon contains three structural genes. One gene codes for lactose permease, which is a membrane protein that facilitates the diffusion of lactose into the cell. Another is for ß-galactosidase, the enzyme that hydrolyses lactose into glucose and galactose. The third enzyme codes for a transacetylase, the function of which is, as yet, unclear.

INDUSTRIAL PRODUCTION OF ENZYMES

BENEFITS OF USING ENZYMES

The **catalytic** properties of enzymes are put to use in many biotechnological applications. As catalysts, enzymes are reused, and relatively small quantities are required within any process. Their **specific** nature means that, when they are added to a fermenter, it is possible to achieve a very precise control of those reactions that take place. This results in fewer wasteful by-products. The relatively low **optimum** temperature of many enzymes can also deliver great cost-savings in many production processes.

INDUSTRIAL USES OF ENZYMES

Paper pulping is an energy-intensive process that is becoming increasingly efficient through the use of enzyme technology.

More than 75% of enzymes used by industry are degradative **hydrolases**, such as protease. Many of these are used in **detergents**, such as biological washing powders. Proteases are used to clean and deodorise wastes of human and animal origin. In medical applications, for example, protease is used to remove the 'bio-burden' from surgical instruments and reusable diagnostic equipment. Other hydrolases include lipase, which is commonly used in restaurants to remove grease from surfaces and drains.

Given the long history of biotechnology in baking and brewing, enzymes are also commonly used in food processing. Amylase breaks down starches to maltose, such as in the conversion of corn starch to syrup. Even more dramatic is the conversion of cellulose to sugars, such as in the upgrading of straw to a more useful animal feed product. The fruit-juicing industry also relies heavily on enzymes (see page 83).

Other sectors that have been revolutionised by the industrial production of enzymes include the textile industry (for example in denim, leather and cotton processing and finishing) and paper-pulp manufacture and recycling (which uses enzymes for bleach-free whitening and de-inking).

SOURCES OF ENZYMES

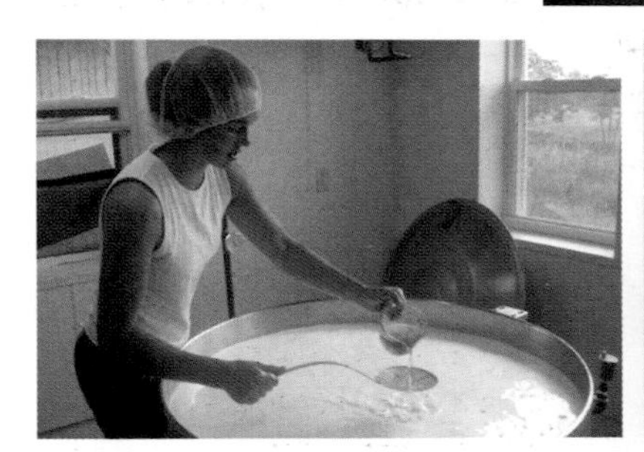

Rennet being added to milk during small-scale cheese manufacture.

In this much larger open fermenter, the addition of the enzyme chymosin has caused the milk protein casein to flocculate and form curds.

Plant tissues and cultures of plant cells are used to produce of hydrolytic amylases and proteases for use in the food industry. Animal tissues and organs are used rarely, with the extraction of rennet (chymosin) from the stomachs of animals for use in cheese manufacture being the only major industrial activity. Instead, microorganisms provide by far the bulk of industrially produced enzymes. Cultures of microorganisms are grown in fermenters to produce naturally occurring enzymes or genetically modified (transgenic) enzymes.

The table below shows enzymes produced industrially using naturally occurring microorganisms.

Enzyme	Source organism	Use of product
cellulase	*Penicillium* (fungus) *Aspergillus* (fungus)	used in washing powders, fabric improvers, fruit juicing and animal feed
pectinase	*Erwinia* (bacterium)	used to increase yield and clarity of fruit juice
amylase	*Bacillus subtilis* (bacterium) *Aspergillus* (fungus)	used in detergents and in food industry to increase maltose sugar content of foodstuffs

ENZYMES PRODUCED USING GENETIC ENGINEERING

Rennet is an enzyme-rich mixture used in cheese production; it is harvested from the fourth stomach (abomasum) of calves. When added to milk, the action of the enzyme **chymosin** splits the milk into solid curd and liquid whey. Genetic engineering has successfully reduced the requirement to use animal rennet; around 80–90% of cheese production in both the UK and USA now uses chymosin produced by genetically modified yeast.

Chymosin acts by catalysing the cleavage of the milk protein κ-casein. The cleavage of this molecule causes the flocculation (clumping) of the hydrophobic casein milk proteins (the curd) and the separation of the hydrophilic constituents (the whey).

Watch an advert for *Chy-Max® chymosin* on YouTube.

A gene for the chymosin protein was generated in the laboratory. This was done by extracting the mRNA for chymosin from a calf abomasum cell, and then using the enzyme reverse transcriptase to generate a suitable cDNA sequence (cDNA is complementary to the RNA).

A plasmid was then cut open with a suitable restriction enzyme and the chymosin gene sealed into the plasmid using ligase. Suitable marker genes on the plasmid allowed the identification of yeast cells containing the recombinant plasmid.

This recombinant plasmid was then inserted into the yeast species *Kluyveromyces lactis*. This eukaryotic cell is able to accept plasmids and is the best choice for commercial production of this particular enzyme.

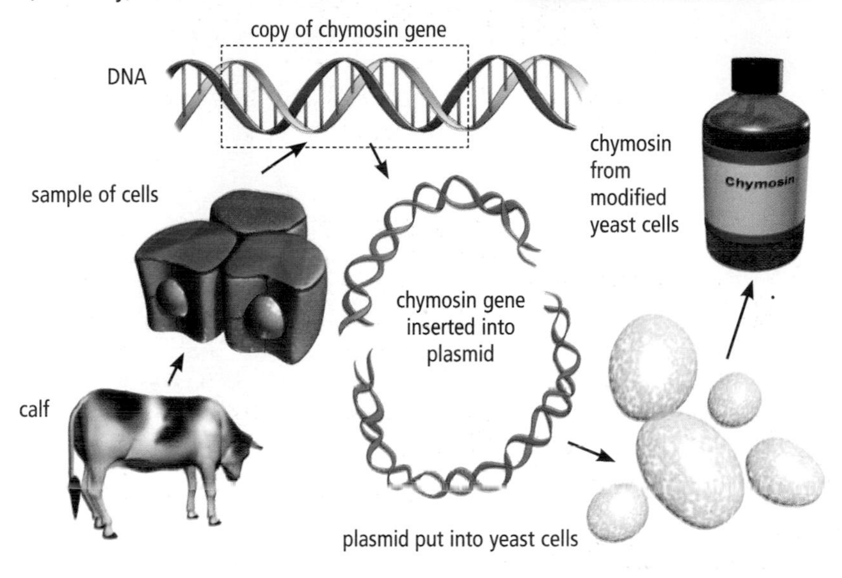

Use of genetic modification and transformation in the production of chymosin

PRODUCTION PROCESS

There are many considerations before enzymes can be produced on an industrial scale. Aseptic technique is critical, and the optimal fermentation conditions in terms of nutrients, pH and time must be established. The practicalities of working with such large volumes of culture medium must be thoroughly understood, such as in terms of the delivery of oxygen, removal of excess heat and need for anti-foaming agents. Careful scaling up is important to avoid expensive mistakes (see page 75).

It is also necessary to ensure that the metabolism of the culture organism is well understood. Some enzymes are synthesised as primary metabolites; that is they are made during the log phase or exponential growth of the culture. Secondary metabolites, on the other hand, are only synthesised during the stationary phase (see page 77). Depending on the enzyme, it may be more efficient to run either a batch or continuous culture. For some secondary metabolites, it may be most productive to use a 'batch-fed' culture. Once the culture is productive, various downsteam processes are required for recovery, concentration and purification.

Stage	Techniques used	Purpose
crude recovery	flocculation, filtration, centrifugation	clumping of colloidial suspension and separation
concentration	ultrafiltration, vacuum extraction	dialysis and evaporation used to remove water and other unwanted molecules
purification	chromatography	capillary chromatography can achieve the extremely high purity required for medicinal, analytical and other applications

Downstream processing in enzyme production

LET'S THINK ABOUT THIS

Biotechnologists may not have a gene available for every useful enzyme. Enzymes can be developed artificially by protein engineering. Rational design involves generating a new primary sequence of amino acids from an entirely artificial DNA sequence. Alternatively, mutagenic agents combined with an artificial selection process can be used to evolve new enzymes.

CELL CULTURE AND AGRICULTURAL APPLICATIONS

ANIMAL CELL CULTURE

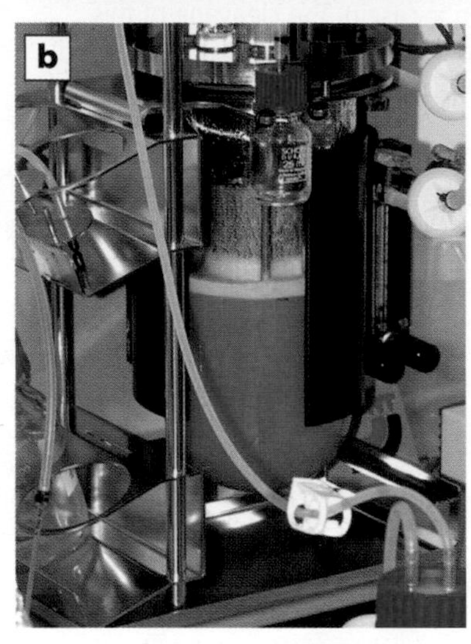

The culture of animal cells requires **aseptic technique**, as well as the control of **growth conditions** (see page 77). The lifetime of **primary cell lines**, cells cultured directly from an animal, tends to be limited, as the cells die after a certain number of mitotic divisions. **Cancer cell lines**, on the other hand, are immortal and can divide repeatedly in culture. Animal cell cultures are used in the production of blood clotting factors and monoclonal antibodies for example (see page 85). Many animal cell culture applications use anchorage-independent cell lines.

(a) In anchorage-dependent animal cell cultures the cells attach to the surfaces of the culture flasks and stop growing when there is a confluent monolayer. (b) In anchorage-independent animal cell cultures the cells grow in suspension in the liquid medium.

PLANT CELL AND TISSUE CULTURE

Some plant cell lines can be grown in suspension culture in a fermenter for production of flavours, colourants, dyes or essences. Although many plant cells grow slowly in culture and are easily damaged in suspension, this field of biotechnology is likely to expand. More typically, plant tissue is grown from **explants** using tissue culture techniques (see page 11).

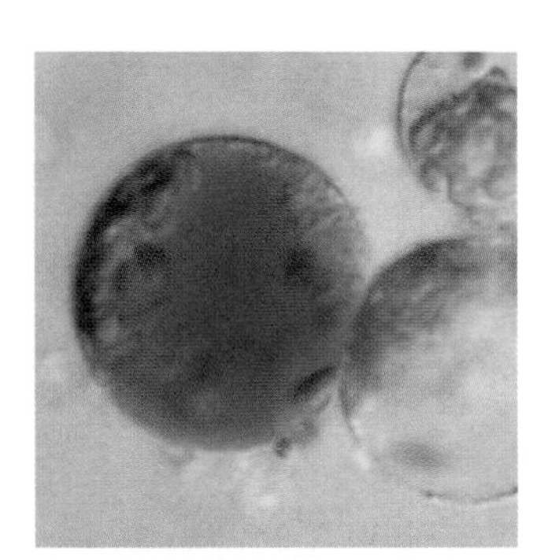

The hybrid cell in this picture was formed by the fusion of a leaf cell with a cell from a red flower petal of a different plant species.

In photosynthetic plant cell and tissue culture, stainless steel fermenters are replaced by transparent materials and **lighting** is provided. For whole-plant propagation light, temperature and humidity must be controlled to produce plantlets.

When plant tissue culture is being used for artificial selection, it is often necessary to remove the cell wall of the explant sample. This is done by digesting the cellulose and pectin using cellulases and pectinases (see page 83) and the cells produced are known as **protoplasts**.

In somatic fusion, the protoplasts of **sexually incompatible** species or varieties can be **hybridised** using **polyethylene glycol** (PEG), a chemical that encourages the fusion of cell membranes. Hybrid cells are allowed to regrow their cell walls and are then propagated. Characteristics of both parental cells, such as high yield and disease resistance, may be displayed in a hybrid. Protoplasts may also be used for genetic modification using *A. tumefaciens* (see page 45).

The domestic banana flower is sterile as the plant is triploid. To protect this important but genetically uniform crop from destruction by viral attack, genetic variation may have to be introduced by protoplast fusion.

DON'T FORGET

The osmotic sensitivity of protoplasts means that they must be formed in an isotonic solution.

For footage of laser microsurgery on plant protoplasts look up www.youtube.com/watch?v=CDCbhDOsZ84

contd

SILAGE PRODUCTION

Silage, preserved grass fodder, is used to feed cows and other grazing animals in winter. The preservation of the nutritional qualities of the grass is achieved by its **anaerobic fermentation**. The grass is **inoculated** with a mixture of enzymes and bacteria. The enzymes include pectinases and cellulases, which partially hydrolyse the cell walls of the grass (see page 83). The bacteria include species of *Enterococcus* and *Lactobacillus*. These bacteria produce lactic acid, which reduces the pH of the silage. Suitably anaerobic conditions can be maintained in several ways. Individual bales can be wrapped in plastic, or larger volumes of grass can be placed in a sealed silo. The anaerobic conditions and low pH prevent the growth of aerobic **spoilage organisms**.

(a) Freshly cut grass is shredded and wilted before collection. This reduces silage run-off during fermentation (see page 69).

(b) The grass is packed into a concrete bunker or metal tower, known as a silo, and air is removed from the heap by compression; an air-tight plastic sheet is spread over the grass for anaerobic fermentation.

ENHANCING NITROGEN FIXATION

The mutualistic relationship between leguminous plants and nitrogen-fixing *Rhizobium* species in root nodules has a beneficial effect on agricultural productivity and yield. Alternative methods of boosting the available nitrogen in soil involve the application of fertilisers. Demand for fertilisers is met by the mining of limited stocks of animal wastes (such as guano in bat caves) or the expensive production of synthetic nitrogen fertiliser, which is costly in terms of the energy required to break the triple bond of nitrogen gas. Nitrogen fixing by bacteria in legumes is cheap in comparison.

There has been much research into the possibility of extending this mutualistic symbiosis to a wider selection of crop plant species. The functions of **nitrogenase** and **leghaemoglobin** have been widely studied (see page 52). Nitrogenase only functions in the microanaerobic conditions provided by leghaemoglobin. There is a high degree of specificity between each species of *Rhizobium* and its host. The specificity is related to *Nod* genes on the bacterial plasmids that allow the bacteria to enter their host cell. The genes for nitrogen fixation, *Nif* genes, are inhibited by high nitrogen concentrations. The genes for leghamoglobin are shared by the two mutualists, with the globin synthesised by the plant and the haem by the bacterium.

The key to enhancing nitrogen fixation through the manipulation of this system lies in controlling expression of these genes and the activity of the proteins produced. Some progress has been made in improving *Rhizobium* strains, but the development of new nitrogen-fixing symbiotic partnerships appears to be some way off.

LET'S THINK ABOUT THIS

Animal cloning using somatic (body) cells may be important in biotechnology in the future. It has been demonstrated with several species, most famously with Dolly the sheep.

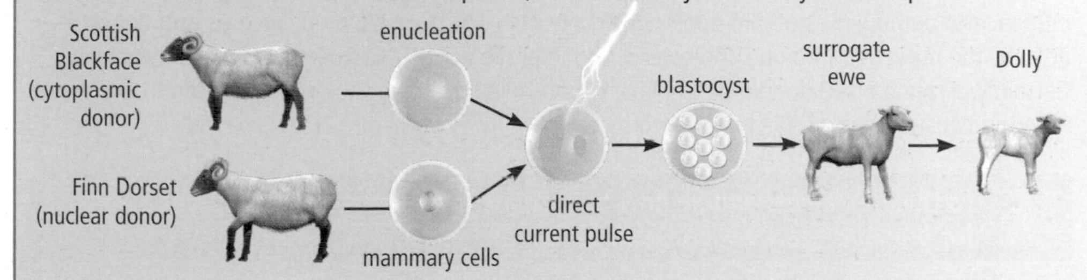

Cloning Dolly – of the four sheep in this diagram, one pair share the same chromosomes, another pair share the same cytoplasm, and yet another pair shared nutrients between a uterus and placenta during pregnancy!

FOOD INDUSTRY BIOTECHNOLOGY

There is a long history of the use of biotechnology in the food industry, with evidence of brewing, baking and cheese-making stretching back for many thousands of years. More recent advances in our understanding of the biology and technology involved have both widened the range of products and increased the reliability of the processes.

FERMENTED DAIRY PRODUCTS

Traditional products

Milk is the only natural source of the carbohydrate **lactose**. The anaerobic bacteria that are found as a **natural inoculant** in milk can ferment this lactose into lactic acid. This lowers the pH of the milk, preventing further **spoilage** by other microorganisms. The increasing acidity (as the pH reduces) denatures the milk's casein proteins. The effect is to curdle the milk, splitting it into solid curds and liquid whey.

Before fermentation, the milk used in fermented dairy products would usually be pasteurised by heating to 72°C for 15 seconds. This 'high temperature–short time' process reduces potential pathogens to below critical levels – microorganism populations are reduced by a factor of 100 000 by the treatment. In this photo, the milk is being cooled rapidly after this pasteurisation stage, preventing unwanted curdling. Pasteurised milk has a refrigerated shelf-life of two to three weeks.

Typically, in the production of fermented dairy products, milk is **pasteurised** to remove the naturally occurring microorganisms. Lactic acid-producing bacteria are added as a starter culture, along with other microbes chosen for the distinctive flavours they add to the finished product.

Find out about Louis Pasteur's contribution to biotechnology at www.epo.org/topics/innovation_and_economy/outstanding_inventors/pasteur.html

In the production of **yogurt**, the milk casein proteins are first denatured by heat. *Lactobacillus bulgaricus* is the lactic acid-producing species and it grows mutualistically with *Streptococcus* which adds to the distinctive yogurt flavour. Other species, such as *L. acidophilus* and *Bifidobacterium* are often used in modern yogurt production.

In **cheese** manufacture, the curdling action of the anaerobic bacteria is enhanced by the addition of rennet (see page 79), before cutting the curd to help release the whey. Cream and cottage cheeses require little further processing. Harder cheeses are made by compressing the curd to remove more of the watery whey. Different cheeses are matured for different lengths of time, which allows different flavours to be developed by the various microorganisms. In blue-veined cheeses, the starter culture includes spores of the aerobic *Penicillium roquefortii*. These only develop once air is deliberately introduced into the cheese later in its production. Cheese with 'holes' is the result of carbon dioxide production by *Propionibacterium*.

Industrial-scale cheese production

Novel probiotic products

Functional foods are those that provide added health benefits beyond basic nutrition. Those functional foods that can aid the prevention or treatment of disease are sometimes called 'nutraceuticals'. Some of the best-known functional foods are live cultures of fermented dairy products. These cultures are designed to be consumed with the aim of reducing the influence of pathogenic gut microbes. Other benefits have been claimed, such as anti-cancer activity, the reduction of blood cholesterol and improvements in lactose intolerance symptoms. Certainly, a randomised double-blind placebo-controlled trial has shown a reduction in diarrhoea with the consumption of this type of product.

DON'T FORGET

'Good bacteria' only exist in adverts!

For help in understanding how to evaluate the various health benefits claimed for products see www.badscience.net

YEAST EXTRACTS

The beer brewing industry produces a regular supply of yeast. Some is sold to distilleries, to replace their yeast culture which is boiled and killed in the distillation process. However, much is available for further processing into yeast extract, a foodstuff rich in B vitamins. **Hydrolysed** extracts are made by boiling the yeast in acid, before neutralising the acid to make a salty product. The **autolysis** of yeast involves heating the cells with salt to encourage the enzymatic self-digestion of the cells. This process produces a number of flavours depending on the enzymes present, the stage of life cycle and age of culture at degradation. Both hydrolysis and autolysis of yeast results in meaty flavours due to amino acids and nucleotides in the extract. The flavours can be used in foods such as crisps.

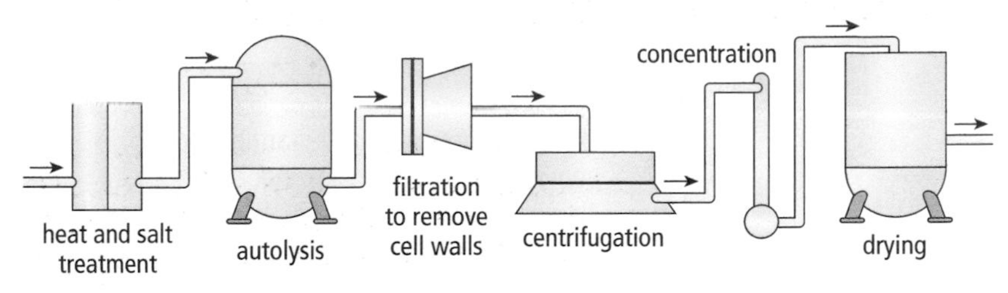

Yeast extract autolysis

FRUIT JUICE PRODUCTS

Plant fruits provide a great food source, but the industrial removal of their juice can be difficult due to the cell walls of the fruit. **Extraction** requires energy and **yield** can be low. A thick product with high **viscosity** and low **clarity** can be caused by a **haze** of complex carbohydrates in the juice. To counteract these problems, **fungal enzymes** such as cellulases, arabanase, pectinases and amylases are frequently used in the juicing industry.

Enzyme	Substrate	Benefit to product
cellulase	cellulose	reduced viscosity
arabinase	arabinan	increased clarity
pectinase	pectin	reduces haze
amylase	starch	increases sugar content

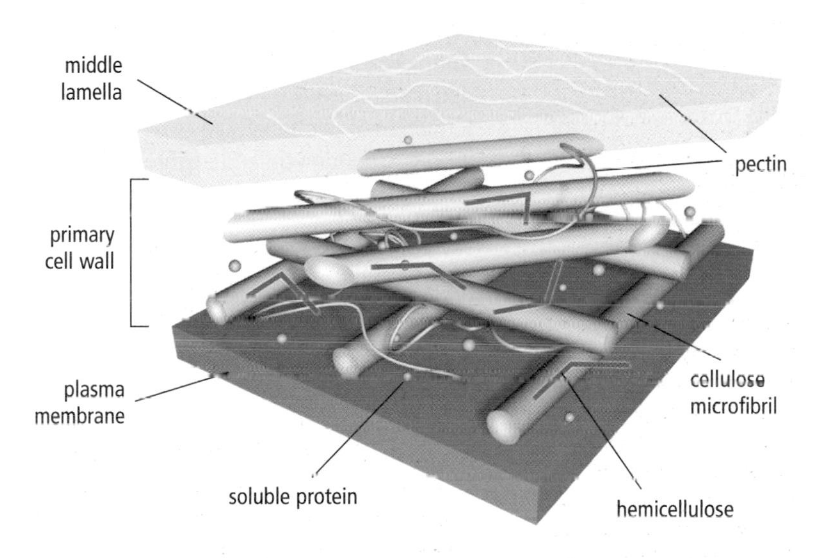

The plant fruit cell wall contains the complex carbohydrates cellulose, pectin and arabinan, all of which can be digested by enzymes to increase juice production.

Search the website www.saps.plantsci.cam.ac.uk for more information on juice extraction experimentation in the laboratory

LET'S THINK ABOUT THIS

Gene silencing is a technology that prevents a gene being expressed. It was used in the genetic modification of the *flavr savr* tomato. In this tomato, the gene that was silenced was polygalacturonase, a pectinase that softens fruit. The aim was the production of fruit with resistance to bruising and increased shelf life. While the latter was achieved, the fruit were more susceptible to bruising than anticipated and the *flavr savr* tomato was withdrawn from the market. Other reasons for its lack of success with consumers were social and economic issues relating to general concerns over the safety, production and use of genetically modified foods. However, gene silencing is not a transgenic technology – it involves the cloning and reversal of a pre-existing gene to produce mRNA complementary to the normal mRNA for the gene. The two complementary RNA strands hybridise and the double-stranded mRNA fails to attach to a ribosome for translation.

MEDICINES

Since the discovery of the antibiotic penicillin in 1929 by Alexander Fleming, biotechnology has had a major role in the development of new medicines and treatments for disease.

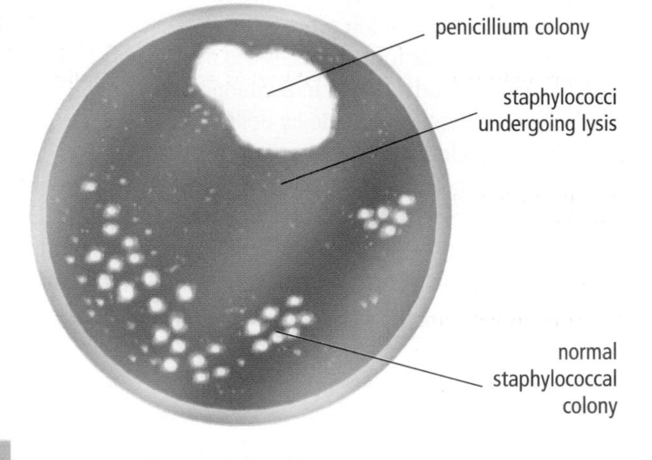

Fleming's famous Petri dish revealed the action of Penicillium *on bacterial colonies for the first time.*

Read about the Petri dish that led to the discovery of penicillin at the *Scottish Science Hall of Fame* www.nls.uk/scientists

ANTIBIOTICS

Antibiotics are chemicals that are produced by microorganisms and inhibit the growth of other microorganisms. For example, penicillin is produced by the fungi *Penicillium* and it inhibits the formation of peptidoglycan cross-links in the bacterial cell wall. Other broad-spectrum antibiotics include streptomycin and erythromycin, which are produced by bacteria.

There are two common modes of action of antibiotics. **Bacteriostatic** antibiotics (such as tetracyclines) interrupt the bacterial cell metabolism, for example DNA replication, so that further growth of the bacterial population is prevented. With **bactericidal** antibiotics (such as penicillin) the bacterial cells are killed by the action of the antibiotic.

The production of penicillin

Various species of the fungus *Penicillium* have been used in the development of industrial-scale production of penicillin. Initially, the fungus would only grow on the surface of static liquid medium and yields were low. Later, different *Penicillium* species with a submerged growth form were discovered.

Modern penicillin production uses a high-yielding mutant form of *Penicillium chrysogenum* in fermenters with a 20 000 litre capacity. The culture medium is often made from **cheap feedstock**, waste materials, starch and plant oils, and tends to include a mixture of glucose (for rapid growth), lactose (to increase yield) and yeast extract (as a source of nitrogen). The ideal growth conditions are monitored and maintained; for example, sterilised air is bubbled through the culture to maintain aerobic conditions and to help control temperature; buffers keep the fermentation at a neutral pH. Control measures, such as sterilisation and appropriate hygiene, are vital to ensure the purity of the fermentation.

Antibiotics are usually **secondary metabolites**. This means that they are only produced in the stationary phase of growth (see page 77). As a result, antibiotic yield is greatest in large low-nutrient cultures. Batches of culture are slowly fed fresh medium and culture is removed for downstream processing of the antibiotic. This processing includes **extraction**, **recovery**, **separation** and **purification** of the product.

DON'T FORGET

Do not confuse the terms antibiotics and antibodies!

MONOCLONAL ANTIBODIES

Nature and production

Antibodies are Y-shaped globular proteins produced by B lymphocytes as part of an immune response. Each B lymphocyte produces one specific antibody that binds to one specific antigen. This binding helps to render an antigen harmless. The spleen and bone marrow are the sites of production of B lymphocytes.

contd

MONOCLONAL ANTIBODIES contd

A serum containing antibodies can be harvested from blood of animals exposed to antigenic material. Some time after exposure, blood is removed and antibodies are extracted. Many different antibodies will have formed to react to different parts of the antigen. Each antibody is made by a single B lymphocyte. A serum with many different antibodies against an antigen is described as **polyclonal**.

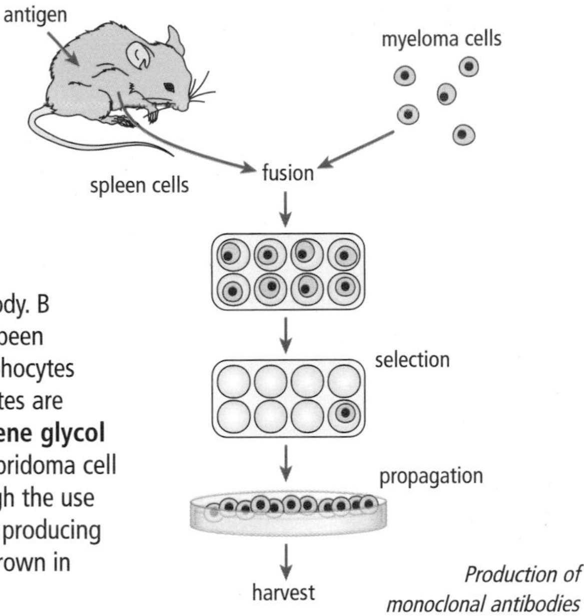

Production of monoclonal antibodies

To produce pure **monoclonal** antibodies, a single line of B lymphocytes must be grown, each secreting the same specific antibody. B lymphocytes can be harvested from the spleen of a mouse that has been exposed to a specific antigen several weeks earlier. However, B lymphocytes do not divide in culture. To get around this problem, the B lymphocytes are **hybridised** with immortal **myeloma** (cancer) cells using **polyethylene glycol** (PEG). The cells produced are called **hybridomas**. Each immortal hybridoma cell line produces only one type of antibody (so it is monoclonal). Through the use of selective media and screening, the hybridoma cell line capable of producing the specific antibody can be identified. Batch cultures can then be grown in fermenters and the secreted antibody can be extracted and purified.

Uses of monoclonal antibodies

Monoclonal antibodies are used in the **diagnosis** and **detection** of disease. For example, **immunoassay** techniques (such as ELISA) involve the use of monoclonal antibodies that have an enzyme attached. This **reporter enzyme** catalyses a colour-change reaction that is used to detect and quantify the presence of a specific antigen. If the antigen is present, the antibody binds allowing the reporter enzyme to produce a coloured product – visual confirmation of antigen's presence. If the antigen is not present, the antibody cannot bind and the enzyme is washed away before any reaction can take place. The antigen can, of course, be another antibody. In this way, blood samples can be screened for either antigens or antibodies for particular pathogens such as AIDS or meningitis. A familiar laboratory demonstration of ELISA involves the detection of the antigens from the fungus *Botrytis*, a pathogen of soft fruit. Pregnancy testing kits use immunoassay technology to detect the presence in urine of human chorionic gonadotrophin (HCG), a hormone only produced by the placenta.

Monoclonal antibodies are also used in the **treatment** of disease. In this technology, tumour-specific monoclonal antibodies are attached to toxins. When the antibodies reach the tumour cells, they combine with their specific antigens, bringing the toxins into close contact with the target cells and ultimately killing them.

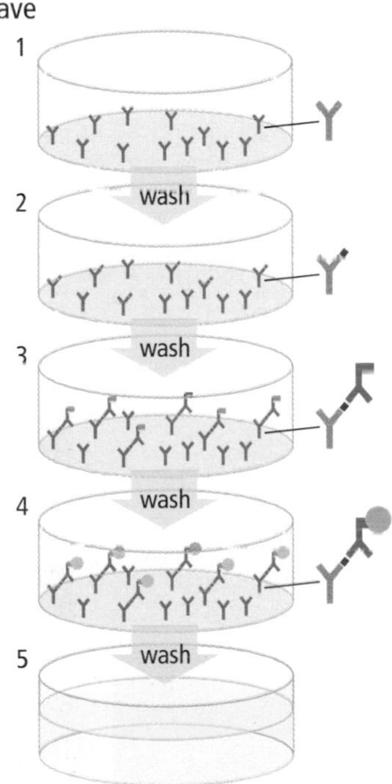

This ELISA involves an antibody (1) attached to the bottom of a plastic well. If a test sample contains the specific antigen then this will bind (2) to the antibody. If an enzyme-linked antibody is added (3) it will bind only where the antigen has bound. If the enzyme substrate is added (4), then a colour-change reaction occurs (5). It is vital that the test well is thoroughly rinsed between each stage.

Use the internet to search for *animations* of an *enzyme-linked immunosorbent assay* and to find out about the use of the monoclonal antibody *trastuzumab* in cancer treatment.

LET'S THINK ABOUT THIS

The misuse of antibiotics in our environment has led to the rapid evolution of antibiotic-resistant strains of bacteria. As a result, MRSA (methicillin-resistant *Staphylococcus aureus*) has become a major problem as a hospital-acquired infection. *Staphylococcus aureus* is a facultative anaerobic bacteria that would normally only cause spots or skin boils in humans. The multiple resistance of MRSA (including to the powerful antibiotic methicillin) can result in severe post-operative infections that may be fatal.

MEASURING BEHAVIOUR

STUDYING ANIMAL BEHAVIOUR

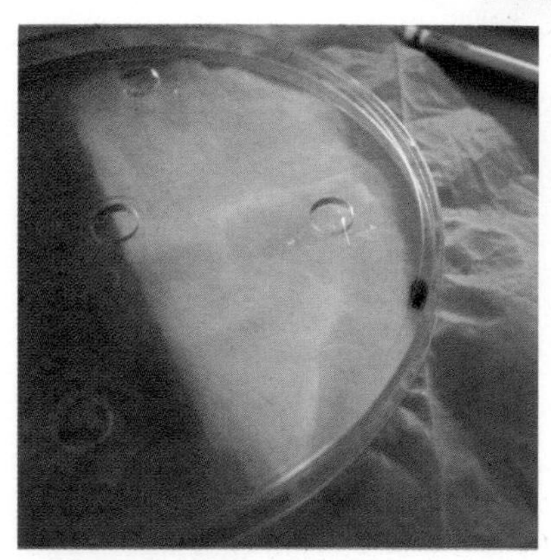

Laboratory studies of animal behaviour, such as choice chamber experiments using woodlice, deliberately reduce the number of possible stimuli to allow clear correlations or causations to be identified.

Behaviour is defined as the observable response that an organism makes to an internal or external stimulus.

In school settings, animal behaviour experiments are often restricted to invertebrate organisms, such as woodlice, and tend to reveal predictable responses to particular stimuli under laboratory conditions. Additionally, attempts are also made to quantify certain aspects of human behaviour, such as in demonstrating learning curves.
At Advanced Higher level, it is important to recognise that behaviour is an **extremely complex adaptation**, and that its study involves variables which can be very difficult to identify, measure or control. The inherent variability in many behavioural responses means that careful experimental design has to be at the heart of this branch of biology. This variability also means that statistical methods are often used to highlight the significance or otherwise of experimental findings.

To understand complex processes, scientists tend to use a **reductionist** approach. In behaviour, this means the classification of complexity into more quantifiable and clearly defined categories.

Events and states

Behavioural responses can be classified as events or states. **Events** are those that are short and discrete movements. For example, sneezing is an event. In contrast, some behaviours occur for long periods of time, and these are known as **states**. Sleeping is an example of a behavioural state. Clearly, organisms are capable of showing more than one behavioural response at the same time, though some are mutually exclusive.

Ethograms

To study the behaviour of a species, it is often necessary to start by compiling an **ethogram** – a list that clearly defines and distinguishes all the different relevant behaviours shown by the study species.

See the University of Strathclyde website at www.strath.ac.uk/biology/environmentenrichment/ethogramconstruction

GATHERING DATA

An Arctic tern chick begs for a sand eel from one of its parents while being brooded by the other. Data regarding complex behaviours like these can be gathered using simple measures such as latency, frequency, duration and intensity.

Forms of quantitative data

Scientific analysis requires data that have been gathered in a **factual** and **objective** manner. The following categories of data are commonly used to quantify animal behaviour.

- **Latency** – the interval of time between a stimulus and its response; for example, the time interval between a chick begging and the chick-feeding response of the adult.
- **Frequency** – how often a behaviour occurs; for example, the number of times a chick begs per minute.
- **Duration** – the length of time for which a behavioural event or state occurs; for example, how long the chick's begging behaviour lasts.
- **Intensity** – a measure of the magnitude, or energy content, of the behavioural response; for example, the volume of the chick's begging calls.

Anthropomorphism

Anthropomorphism is the attribution of human motivation, characteristics or behaviour to non-human animals. The use of anthropomorphism to draw conclusions about or to explain animal motivation and behaviour should be avoided. For example, after consuming food a chick may shake its head. Without further evidence, it would be a mistake to conclude that this indicated that the bird had not enjoyed eating the food.

contd

GATHERING DATA contd

Minimising observer bias

The responses of study organisms to the presence of a human observer will confound the validity of an animal behaviour study. Remote data gathering technology of various kinds (satellite, radio signal, photography) has revolutionised the types of study that can now be carried out without direct observer influence on the study species. The analysis of recorded data also has the advantage that it can be checked repeatedly by independent observers.

Understanding behaviour is essential in the management of wild species, such as in the conservation of the Iberian lynx – the world's most endangered species of cat. Data gathering on this extremely elusive predator now uses a combination of traditional tracking techniques (a), alongside remote data-gathering technology such as the use of a radio collar and video recording (b).

Find out about the satellite tracking of migratory Scottish ospreys at www.roydennis.org

THE CAUSES OF BEHAVIOUR

Behaviour studies often try to answer the question 'why does this behaviour occur?' Such a question can have both a **proximate** and an **ultimate** answer.

Proximate causes of behaviour

Proximate causes include the **trigger stimuli** for the behaviour. For example, in nest building by birds in spring, the trigger is a change in photoperiod – so a change in photoperiod is a proximate cause of birds' nest-building behaviour. Other proximate causes include the hormonal and nervous pathways that induce the response.

Ultimate causes of behaviour

Ultimate causes are the underlying evolutionary reasons for the behaviour. For example, using the same nest-building example, the evolutionary reason for the behaviour is to **improve the survival** of young. The drive to pass copies of genes on to future generations is an ultimate cause of building nests.

LET'S THINK ABOUT THIS

When trying to understand an animal's behaviour it is helpful to consider the following chronology of influences.

1. **Ancestry**: the evolutionary history of the species and genotype will determine an individual's inherited predisposition for innate aspects of behaviour.
2. **Development**: behaviour changes as individuals mature and the influence of learning modifies behaviour.
3. **Trigger stimuli**: proximate stimuli trigger hormonal and neurological cascades that control behaviour.
4. **Function**: successful behaviours are those that ultimately increase survival of offspring (or offspring of relatives).

DEVELOPMENT OF BEHAVIOUR

NATURE AND NURTURE

In the study of animal behaviour, formerly much emphasis was placed on distinguishing between the influences of nature versus those of nurture on the development of behaviour. The role of nurture (learning) was championed by **experimental psychologists** such as B.F. Skinner. The role of nature (genetic or innate influences) was favoured by **ethologists** such as Nico Tinbergen, Konrad Lorenz and Karl von Frisch. Currently, a more balanced view recognises that an animal's behaviour tends to be controlled by both innate and learned influences.

Innate behaviours

Innate behaviours are those that are under strong **genetic control** and tend to result in a **stereotypical** response. This means that they show little variation within or between individuals. Innate responses are often of clear adaptive significance in terms of survival or reproduction.

A young green turtle shows innate behaviour when it digs itself out of its nest and, within hours of hatching, makes its way to the sea.

Learned behaviours

Learning is the modification of behaviour by experience. Learned behaviours are characterised by their variability – individuals have different experience and, for this reason, may have learned to respond differently to the same stimulus. Research in various species has shown that the ability to learn is both genetically determined and under environmental influence. The degree to which learning is important for a species is related to aspects of the species' ecology – for example, the long lifespan and provision of parental care in **primates** gives both the time and opportunity for learning; a shorter lifespan and lack of overlap in generations (preventing parental care of offspring), reduces the opportunity for learning in many **invertebrate species**.

EXAMPLES OF THE DEVELOPMENT OF BEHAVIOUR

An adult reed warbler feeds a young cuckoo. The reed warbler is a species with a complex song pattern, which indicates an extended sensitive phase for song learning. Its song can even include the mimicry of other species' songs. The cuckoo, on the other hand, is a brood parasite and does not have the opportunity to interact with adult members of its own species as a chick – its simple, well-known and stereotypical song is almost certainly innate.

The development of bird song

Experimental studies on the development of bird song have shown that song birds have a **sensitive phase** or **critical period** when the learning of the specific song of the species can occur. The correct development of the song depends on the young bird hearing a song that matches an innate template during this sensitive phase. Those without the opportunity to learn, such as in chicks deafened experimentally, birds only learn to sing a rudimentary 'template' version of their species' song.

Worker honey bee behaviour

Studies on honey bees have demonstrated how innate influences can result in stereotypical yet complex behaviour patterns. Honey bees are social insects; haploid worker bees cooperate to raise the offspring of their queen. Each worker bee lives for up to about 40 days. In this time their behaviour changes from cell cleaning (days 1–2), to nursing larvae (3–11), wax production (12–17), entrance guarding (18–21) and foraging (22–42).

Foragers collect pollen and nectar for the colony and communicate to other workers when good sources of food are located. If the source is close to the colony, the forager will perform a **round dance** on its return to the hive. If the good source of food is found at a distance of more than 80 m, the successful forager will communicate the flight direction and flight duration to other workers using an elaborate but innate **waggle dance**.

contd

EXAMPLES OF THE DEVELOPMENT OF BEHAVIOUR contd

Imprinting

Imprinting is the well-known process by which a young animal, such as a game bird, develops a preference for its mother. It is defined as an **irreversible** learning process with an adaptive significance in terms of protection of young or sexual behaviour. A newly-hatched chick has a critical period or sensitive phase during which it forms an **attachment** to the exclusion of others. In a wild situation, this attachment would be to its mother, and the behaviour would increase the **survival** chances of the young; in artificial situations, objects, people and other animals have been the subjects of imprinting. Imprinting while young can also influence **mate choice** later in life.

SIGN STIMULI AND FIXED-ACTION PATTERNS

When individuals of the same species interact, a **sign stimulus** or releaser from one individual often elicits a stereotypical **fixed-action pattern** response from another. The response is innate, so once initiated it runs to completion. For example, a young herring gull chick responds by pecking the sign stimulus of a red spot on the adult beak. This pecking acts as a stimulus to the adult for the fixed-action pattern of feeding the chick. A series of sign stimuli and fixed-action pattern responses can produce complex behaviours, as illustrated by stickleback courtship (see table below).

Sign stimulus	Fixed-action pattern
swollen belly (eggs) of female	zig-zag dance by male
zig-zag dance by male	following by female
following by female	indicating nest entrance by male
indicating nest entrance by male	entering nest by female
entering nest by female	trembling by male
trembling by male	egg laying by female
egg laying by female	release of sperm by male
non-swollen belly of female	aggressive behaviour by male

The development of the red belly of the male stickleback is a response to the stimulus of increasing photoperiod. A series of sign stimuli and fixed-action patterns occur during stickleback courtship.

LET'S THINK ABOUT THIS

Fixed-action patterns are clearly valuable in terms of both survival and reproduction. However, a disadvantage is that they can be hijacked by other species. The bee orchid pictured here mimics the sign stimulus of a species of queen bee. The drone bees (males) are duped by the flowers into mating with them, the act of mating being a fixed-action pattern. Mating with flowers is, of course, of no benefit to the drone. On the other hand, the flowers are adapted to transfer pollen onto the drone while he is copulating. If the drone is attracted to several flowers of this species, he will pollinate them.

EVOLUTION OF BEHAVIOUR

NATURAL SELECTION OF BEHAVIOUR PATTERNS

Evolution occurs when genetic variation determines phenotype variation and there is selection of the **best adapted** (the fittest) individuals.

Selection of those best adapted to survive in their environment is termed **natural selection**. When the selection is of those best adapted to breed successfully, this is termed **sexual selection** (see pages 94–95).

Like any other aspect of an organism's phenotype, behavioural characteristics are subject to selection pressures. Behaviour has an inherited (innate) component that is determined by genes and these genes may show genetic variation. Some behaviours result in greater survival or reproductive success, so that these favourable adaptations accumulate in populations.

Learned behaviours, which are not under genetic control, can also be passed culturally from generation to generation. (Evolutionary biologists have coined the term 'memes' for this form of non-genetic inheritance; cultural practices could be considered memes, for example.)

An example of the evolution of behaviour

One example of evolutionary change in a behaviour involves the blackcap, a small migratory species of warbler. Traditionally, this bird is a summer visitor to Scotland, migrating in a south-westerly direction in autumn to its wintering grounds in Spain. Recently, blackcaps have started wintering in Scotland. Ringing studies have shown that these birds are from a population that breeds in central Europe and they have recently evolved a north-westerly migratory direction in autumn. In the laboratory, F_1 hybrids between these two populations have been shown to possess an innate migratory direction intermediate between the two parental populations.

A male blackcap is trapped in a mist net for scientific ringing. A numbered metal ring will be placed on a leg to act as a unique identifier should the same bird be retrapped elsewhere. Coloured rings can also be used to allow identification of individuals without the need for retrapping as shown by the reed warbler on page 88.

Look up scienceblogs.com/notrocketscience/2009/12/british_birdfeeders_split_blackcaps_into_two_genetically_dis.ph

Extended phenotype

The influence of an individual's genotype on the physical characteristics of its phenotype is well understood. The genotype can also influence behavioural characteristics, and this is known as the **extended phenotype** – where **genes have an influence outside the body** of an organism. Common examples include the nest-building behaviours of birds, and shoaling and herding behaviour in fish and mammals.

These pictures show the extended phenotypes of herring gull and kittiwake in terms of nesting behaviour. (a) The herring gull builds a simple nest on a relatively horizontal cliff top. The young are cryptically camouflaged and will run and hide at the approach of a predator. When an adult arrives at the nest, the young peck at the red spot on the adult's beak to encourage feeding. The kittiwake (b) builds its nest on a vertical cliff, gluing the structure together with its droppings. The young move very little in the nest and crouch if danger approaches. The kittiwake chicks also move very little during feeding – the adult lacks a red spot on its beak but instead has a red tongue which it reveals to the young to signal the initiation of feeding.

contd

NATURAL SELECTION OF BEHAVIOUR PATTERNS contd

Herding behaviour is part of the extended phenotype in African herbivorous mammals. In general, herding as a response to predators is found most commonly in the larger-bodied grazing species living in the most open habitats. The extended phenotype of the smallest-bodied antelopes tends to involve living alone or in small family groups, feeding by selective browsing, and 'freezing' in response to a predator threat.

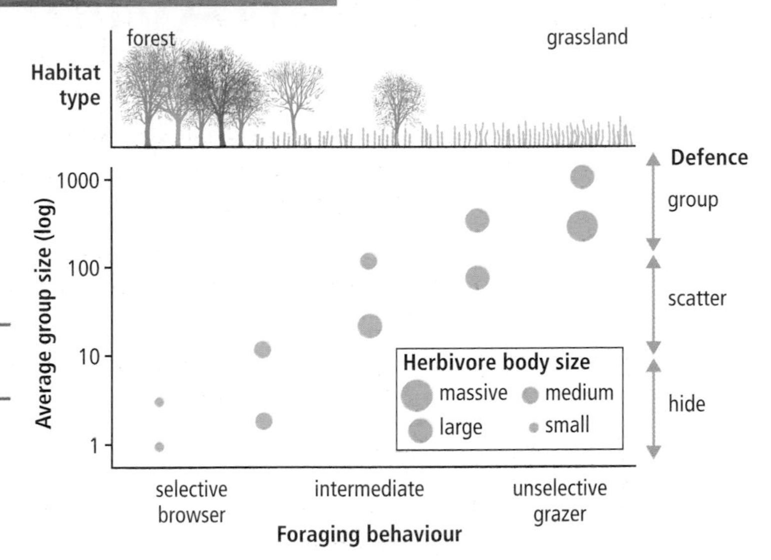

SINGLE GENE EFFECTS ON BEHAVIOUR

Behavioural geneticists have sought examples of behaviours controlled by a single gene (to study in the same manner as other genes in a classic monohybrid cross). The *per* gene in the fruit fly *Drosophila* controls the **circadian rhythm**, the cycle of periods of activity and inactivity in any 24-hour period. Different alleles of the *per* gene have been found that alter the production of the per protein, which in turn has been found to affect the number of hours the *Drosophila* take to complete a daily cycle.

DON'T FORGET

An allele is an alternative version of a gene.

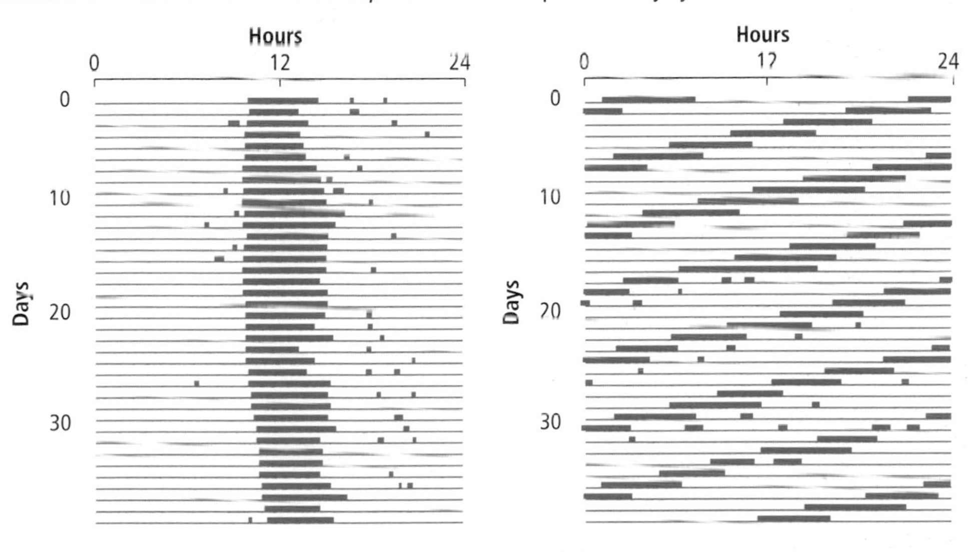

Mutation of the tau *gene in hamsters has a similar effect to the* per *gene in flies. On the left is the normal circadian pattern of a non-mutant hamster; on the right is the reduced circadian rhythm of a homozygous mutant. The active period, shown in blue, occurs broadly at the same time every 24 hours in the normal hamster. In the mutant, the daily cycle is approximately 20 hours in duration.*

BEHAVIOURAL ADAPTATION TO HUMAN INFLUENCE

The impact of humans on the environment often results in rapid changes that reduce the complexity of habitats. Specialist species, such as the giant panda, are unable to cope with rapid change. **Generalist species**, those that are innately adaptable, are able to modify their **habitat preference**, **diet** and **foraging behaviour**. Examples of successful generalist species include the herring gull and the fox. These species now commonly live and breed in urban areas, eating refuse and scraps. Rightly or wrongly, both are considered pest species in some urban areas as a result of their adaptability.

LET'S THINK ABOUT THIS

At one point it was considered that many individual behaviours were under the control of individual genes. As genome studies have advanced, the idea of every behaviour being controlled by its own gene has been abandoned – we simply do not have enough genes for this. Instead, complex cascades of gene products are thought to be responsible for the genetic control of behaviour. We can no longer use genes as an excuse for bad behaviour!

FEEDING BEHAVIOUR

FORAGING AND FEEDING

Foraging refers to behaviour that is related to searching for and finding food. Feeding refers to the manipulation and ingestion of food. Organisms tend to forage optimally – that is to **optimise** the **net** energy gain from foraging and feeding. Energy gain is the total energy intake minus the total energy costs.

$$\text{net energy gain} = \frac{\text{energy intake}}{\text{associated with food}} - \frac{\text{energy expenditure associated}}{\text{with foraging and feeding}}$$

The costs associated with foraging increase if prey is rare or difficult to locate as the **encounter rate** will be low. Similarly, if the prey is difficult to kill or prepare for ingestion, the **handling time** increases, which also increases the costs of feeding.

The gain associated with feeding tends to be measured in terms of energy gain, although, other nutrients are also important.

Optimal foraging in the white wagtail. The prey sizes selected by the wagtail do not reflect how commonly available those prey sizes are. Instead, prey selection is commonest for prey sizes that have the highest net energy gain. Generally larger prey sizes, like this damselfly, are avoided if possible as they incur handling-time costs.

Optimal territory size in birds

Territoriality in birds is an adaptation to reduce intraspecific competition. The benefits of a territory can be measured in terms of the energy content its resources can provide. The energy costs of a territory are those required in its defence against competitors.

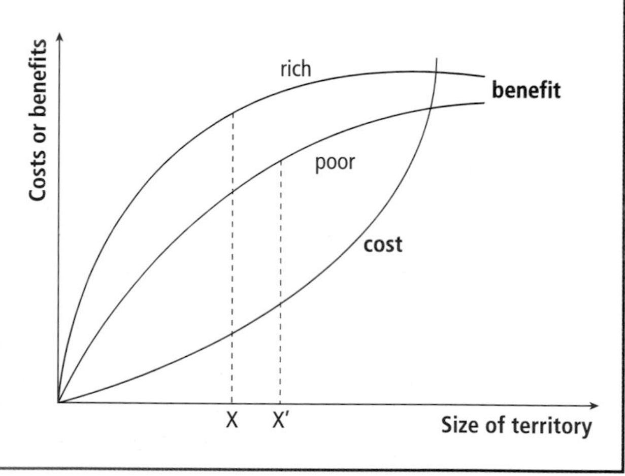

Optimal territory sizes (X and X′) in two different habitats. Territories have to be larger in the habitat with poor resource benefits compared to the rich resource habitat. The cost of defending territories increases with their size.

DON'T FORGET

Optimum does not mean maximum – it means *net* maximum.

COMPARISON OF SOLITARY AND COOPERATIVE HUNTING

An understanding of optimal foraging sheds light on the evolution of cooperative hunting. By hunting cooperatively, some species of predators are able to increase both their hunting success rate and the size of the prey that they target. Examples of cooperative hunters include grey wolf, lion, African wild dog, various dolphins and the Harris hawk. Other predatory species, such as the Iberian lynx, forage optimally as solitary hunters.

African wild dogs hunt cooperatively. In a hunt the pack tends to run in single file. The leading dog (usually the dominant male or female) selects a target and pursues it. The pack chases at a speed that is too slow to catch an antelope in a short chase, but the dogs have the advantage of greater stamina. If the prey runs in a curve, as territorial species often do, the trailing members of the pack cut the corner and either bring the prey down or bite and slash it to allow the other dogs to catch up. Some packs have learned particular behaviour patterns, such as immobilising a large prey item by biting its nose, and these memes (see page 90) are passed on culturally.

DEFENCE STRATEGIES

Mimicry

Behavioural mimicry is commonly used to confuse or dupe a potential predator. For example, a bird may feign injury to lure a predator from its nest. Batesian mimics may move in the manner of the harmful model species (see page 57). For example, the clearwing moth resembles a wasp by flying during the day. (Moths are normally nocturnal.)

Camouflage

Crypsis is a form of camouflage whereby colouration and markings match closely to the background. Organisms that show cryptic camouflage must carefully select their background, and then generally remain still for periods of time.

Masquerade describes a form of camouflage whereby an organism appears to be another object. For example, several species of birds resemble broken branches and sit in an appropriate position. Several species of caterpillars resemble bird droppings and tend to sit on the upper side rather than underneath leaves.

Disruptive colouration breaks up the outline of prey species so that individuals are less easily recognised by a predator. This colouration may be associated with unusual movements in some species.

Vigilance

Prey species must keep watch for predators to increase their chances of survival. This predator-scanning behaviour is known as **vigilance**. Solitary animals spend more time being vigilant than members of groups. Species that live in groups benefit from an increased total group vigilance which reduces the individual need for vigilance. Of course, species that live in groups suffer from the fact that groups of prey tend to attract more predators.

Escape response

What happens if the above defence strategies fail and the prey is detected? Many species also have an escape response – burrowing, running, jumping, flying or swimming away.

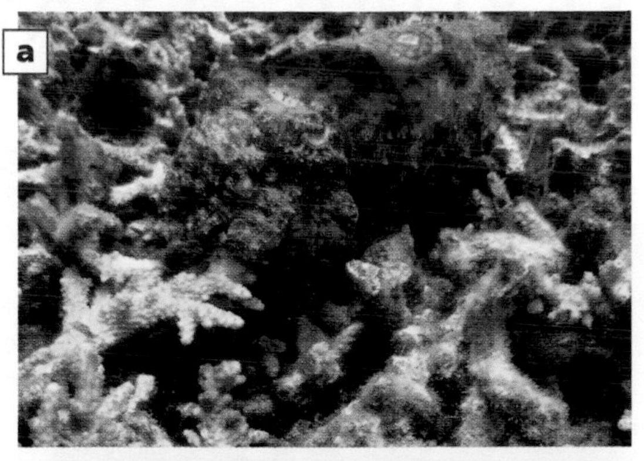

Cuttlefish showing (a) crypsis, (b) masquerade, (c) disruptive colouration and (d) a jet-propelled escape response.

LET'S THINK ABOUT THIS

An evolutionary arms race can develop between predators and their prey. Predators provide the selection pressure for the survival of increasingly better adapted prey. Prey defences provide the selection pressure for the survival of increasingly better adapted predators. Explore this evolutionary arms race using the example of the rough-skinned newt and the common garter snake (http://evolution.berkeley.edu/evolibrary/article/biowarfare_01).

SEXUAL BEHAVIOUR

MALE AND FEMALE INVESTMENT

As sessile animals unable to move location, these coral colonies and this clam are restricted to external fertilisation. Each species tends to evolve a synchronised release of gametes to maximise the chances of successful fertilisation.

Sexual reproduction requires the coordination of behaviour between individuals of different genders. This can mean the simultaneous release of gametes into an aquatic environment in species with **external fertilisation**, or the sometimes very lengthy courtship that precedes **internal fertilisation**. The latter increases the chances of gametes meeting and fusing successfully, so fewer gametes are required. This, therefore, reduces the total costs of gamete production. Internal fertilisation also offers increased protection for the gametes, both from predation or abiotic threats. Almost all terrestrial organisms use internal fertilisation.

There are costs associated with reproduction and these are known as **parental investment**. These costs of parental investment are not often shared equally between the sexes and they are usually greater in females. The inequality arises during gamete production – males produce many small gametes each with a very low individual cost; females produce fewer but larger gametes, each of which has a greater energy store and, therefore, a greater energy cost. Parental investment does not stop at gamete production, of course. The costs may also include energy and resources required for courtship and mating, embryonic protection and incubation, and the nurturing of young.

REPRODUCTIVE STRATEGIES

As with foraging, optimality is important in the evolution of sexual behaviour. Organisms attempt to optimise both the *number* and *quality* of the offspring they produce, as in doing so they maximise the number of their genes that are passed on to future generations. In this way, the costs of any current parental investment are always balanced against the production and survival of current *and future* offspring.

Two different reproductive strategies that have evolved are monogamy and polygamy. In **monogamy** an individual has only one mate during its breeding season or lifetime. In **polygamy** an individual mates with several others during one season. In addition, the different genders may have different optimal reproductive strategies.

COURTSHIP AND DISPLAY

Darwin recognised that any characteristic which increases the breeding success of one gender will tend to become more common in successive generations of that gender, and he termed this **sexual selection**. As a result of sexual selection, different adaptations tend to evolve in the two genders – the characteristics that increase the likelihood of reproductive success of males may not be useful adaptations in females and *vice versa*. As a result, the sexes of many species differ in both appearance and behaviour. Differences between the two sexes are described as **sexual dimorphism**.

The sexual dimorphism in giraffes is most obvious during courtship; the male is the taller of the two sexes.

contd

COURTSHIP AND DISPLAY contd

Given the costs and potential benefits associated with reproduction, sexual selection has resulted in some complex and lengthy courtship behaviours involving very elaborate display.

In some species success in **male–male rivalry** can increase the access of males to females. The successful males will be those that are stronger or have greater weaponry, and these are fitness characteristics that will be advantageous to offspring.

In other species, **female choice** drives the evolution of conspicuous markings, structures or behaviours in males. These male displays tend to reveal 'honest signals' – characteristics that allow females to assess the genetic quality of the males. Good phenotypic quality is an indicator of 'good genes'.

Male–male rivalry results in the sexually dimorphic evolution of weaponry in males; (a) Spanish ibex and (b) impala. In both these species, the females lack horns and also have a smaller body size.

A cryptically plumaged female of the sexually dimorphic Himalayan monal pheasant assesses the conspicuous and colourful plumage of the male. In the wild, a female would try to visit several males in order to choose a mate with the best genes to pass on to her offspring.

AVOIDANCE OF INBREEDING

Inbreeding is the sexual reproduction of close relatives or of individuals with very similar genomes. Inbreeding tends to lead to highly homozygous genomes and reduces the genetic variability in a population. This can be disadvantageous where harmful recessive alleles are present in the population. Individuals that are homozygous for these alleles tend to have reduced fitness and reduced breeding success.

To reduce the chances of inbreeding, the two genders often show different dispersal behaviour patterns. For example, in social mammals, the young males disperse, whereas in birds the young females move away from their natal area (the area where they were raised). Both of these dispersal patterns reduce the chances of siblings or other close relatives breeding together in the next generation.

DON'T FORGET

Inbreeding does not cause mutations, but it does increase the chance of offspring being homozygous for recessive alleles.

Look up www.plosone.org/article/info:doi/10.1371/journal.pone.0005174 to find out about the role of inbreeding in humans in the extinction of a European royal dynasty.

LET'S THINK ABOUT THIS

When a population first becomes inbred it is usually the heterogametic sex (the sex with the non-homologous sex chromosomes) that is most affected in terms of fitness. Is it a coincidence that it is the heterogametic sex that is most likely to disperse in both mammals (the males are XY rather than XX) and birds (the females are ZW rather than ZZ)?

SOCIAL BEHAVIOUR

A hippopotamus social group consists of a dominant bull and a group of females and young. The bull will defend the territory from other hippo groups to reduce intraspecific competition.

Social behaviour has both costs and benefits. For example, living in close contact with members of the same species risks the rapid spread of parasites. On the other hand, a social group may be safer from predators through group-defence strategies. The benefits of group membership are related to increased survival due to increased protection, opportunities for cooperative hunting and division of labour.

Various ecological factors determine whether social behaviour evolves in a particular species. The patchiness of resources may be a key selection pressure; for example, if nesting sites are restricted to a limited habitat (for example, seabird colonies) or if prey is large or distribution clumped, then social behaviour may be more likely to evolve.

AGONISTIC AND APPEASEMENT BEHAVIOUR

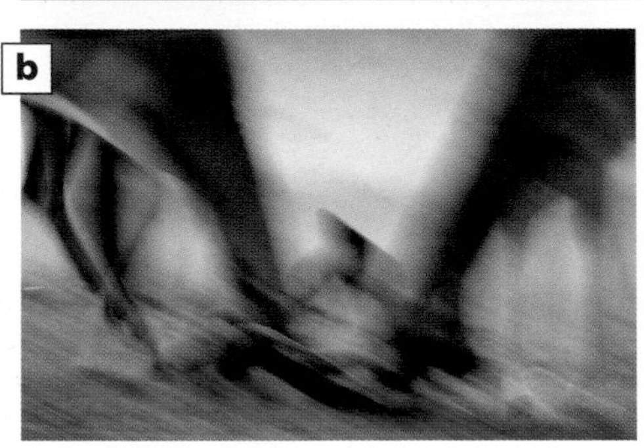

Agonistic behaviours in the impala. (a) Ritualistic display and appeasement; with ears held forward in an appeasement gesture, a ram sniffs the glandular areas of the forehead of a more dominant male. (b) In situations where dominance and submission cannot be established by ritualistic display and appeasement, fighting can occur; this type of agonistic behaviour can be fatal.

Living in groups increases the frequency of interactions between individuals. Without social behavioural adaptations, group living has the potential to lead to repeated conflict. All behaviours associated with conflict, such as fighting or fleeing, are known as **agonistic behaviours**. Social species tend to evolve **ritualistic display** to signal threat, and **appeasement behaviours** to signify submission. These behaviours usually replace actual attack and defence, and both therefore reduce the risk of injury and reduce the energy expenditure involved in resolving conflict.

DON'T FORGET

Agonistic behaviour includes all behaviour associated with conflict, not just aggressive behaviour.

SELFISH AND ALTRUISTIC BEHAVIOUR

The selfish gene

Genes that are good at increasing the number of copies of themselves are the genes that survive through natural selection. So, in the long term, successful genes are the ones that are able to assist the survival of organisms (and their relatives that share these genes). The term 'selfish gene' is used to describe the apparent self-preserving nature of successful genes.

Altruism

Altruism describes behaviours that decrease an individual's chance of survival but increase the chances of survival of others. Altruism is unlikely to evolve unless the benefits of selflessness outweigh the costs. For example, in social species the costs of helping another organism are outweighed if the helper is repaid at a later date – this is known as **reciprocal altruism**. In a reciprocal strategy, cooperators benefit as they tend to have a higher reproductive success; the genes for altruism are then more likely to be passed on.

Kin selection

As close relatives share so many alleles, natural selection often favours behaviours that increase the survival of relatives. In evolutionary terms, this is known as kin selection. One example would be where an organism is better off helping to raise the offspring of a close relative, rather than trying and failing to reproduce themselves.

SELFISH AND ALTRUISTIC BEHAVIOUR contd

Hamilton's rule states that helping relatives is beneficial when:

$$rb - c > 0$$

That is, when the relatedness coefficient (r) times the benefit of helping the relative (b) is greater than the cost of helping the relative (c). The coefficient of relatedness is the proportion of shared alleles (that is, in siblings r = 50%, in offspring r = 50%, in grand-offspring r = 25%, and so on).

Kin selection appears to be important in the evolution of social behaviour. The social organisation of many species is often based around groups of same-sex siblings. For example, a pride of lions is often composed of groups of related lionesses and either an unrelated male or a group of separately related males. This high degree of relatedness between adult members of the group and every offspring can mean that the survival of any offspring is advantageous to any adult member of the group.

SOCIAL ORGANISATION IN PRIMATES

Social organisation in many species of primates increases the chances of group members surviving, obtaining food and reproducing successfully. The long period of parental care in primates give an opportunity for the learning of complex social behaviours (see page 88).

In primates a **social hierarchy** is established by threat and display. This hierarchy is a system of **social ranking. Dominance–subordinance** determines access to food, shelter and mates.

Social behaviours in primates. (a) The high rank of this male vervet monkey is signalled by the intense colours of his genitals; in this ritualistic display he is also revealing canine teeth. (b) The long period of parental care allows plenty of opportunity for learning for this young proboscis monkey. (c) Sexual presentation and alliance formation in the chacma baboon; both baboons are being vigilant to ensure that no dominant baboons are watching them before mating.

Displays and appeasement are used to reduce the chance of injury and wasted energy. Typical examples of **ritualistic display** behaviour by dominant individuals include a baboon displaying its canine teeth or chest beating in the gorilla. This sort of display allows an opponent to assess the threat. If an opponent does not want the conflict to escalate, then **appeasement behaviours** – such as specific facial expressions, submissive body postures, grooming and sexual presentation – are used.

In some species of primates, alliances (bonds) form between individuals. These alliances, which are often strengthened by grooming, are important in maintaining or improving position within the hierarchy.

Find out about the strange lifestyle of the naked mole rat at www.bio.davidson.edu/people/vecase/Behavior/Spring2004/lyons/Reproduction.html

LET'S THINK ABOUT THIS

The evolution of humans is well understood. However, humans are unique in their willingness to cooperate in such large groups of unrelated individuals, including in the sharing of resources. The key may lie in an ability to form cooperative alliances among extended social networks. Studies of archaeological and fossil evidence reveal a relatively recent and rapid increase in social group size.

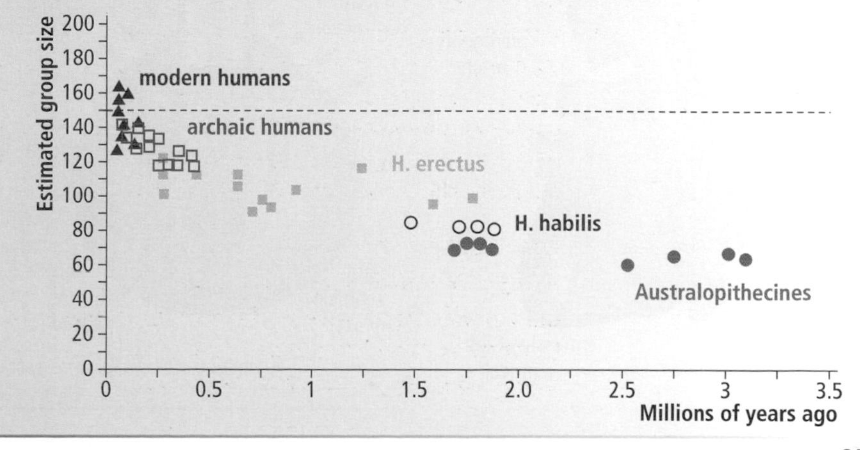

CARDIOVASCULAR SYSTEM (CVS)

COMPONENTS OF THE CVS

The cardiovascular system is made up of the blood vessels and the heart. The heart pumps blood through the vessels. The blood transports many materials around the body:

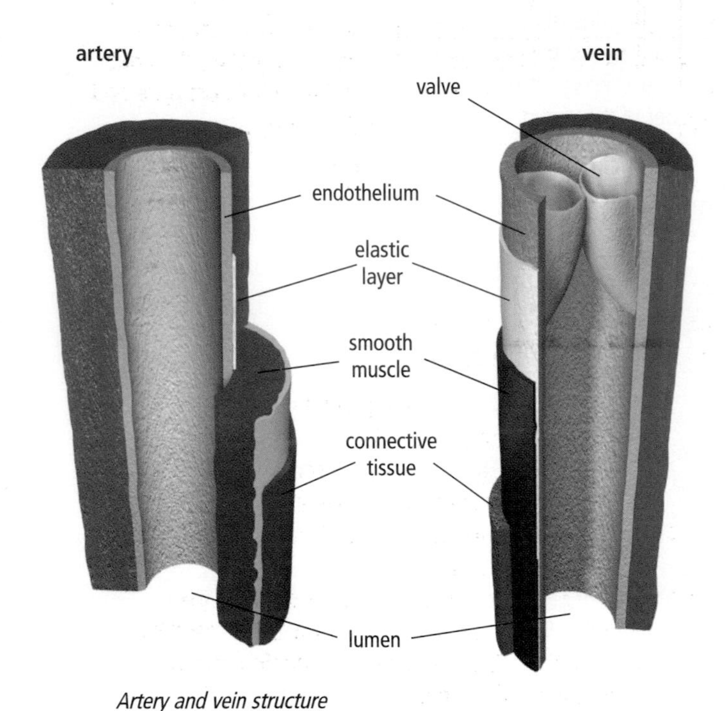

Artery and vein structure

- oxygen to respiring cells and carbon dioxide away from these cells
- digested food from the digestive system to the liver for processing or storage
- urea from the liver to the kidneys
- hormones and antibodies
- it is also important for distributing heat around the body – at rest, most of your body's heat is generated by your liver!

Blood vessels

Blood travels away from the heart in **arteries**. To withstand the high pressure generated by the contraction of the heart muscle, the artery wall has a **thick layer of muscle** and extra **elastic layers**. Arteries divide into smaller arterioles which continue to divide until they become capillaries. A capillary is only wide enough to let red blood cells through in single file. The capillary wall is only one cell thick so that materials can diffuse easily to and from the blood. Capillaries join together again to form small veins called venules. The venules join to form **veins** which carry low pressure blood. The wall of a vein has a **thin layer of muscle** and may include **valves** to prevent blood flowing back.

THE ACTION OF THE HEART

DON'T FORGET

Dr Syc might help you remember that **d**iastole is **r**elaxation, **sy**stole is **c**ontraction of the ventricle muscles.

The **myocardium** is made up of cardiac muscle. This type of muscle has small interconnected cells; these interconnections allow the rapid spread, from cell to cell, of the impulses that cause the contractions of the heart. These impulses are initiated by the sinoatrial node, high in the right atrium. The spread of these impulses is coordinated so that the muscles of the atria contract first to pump blood into the ventricles. This is followed by contraction of the ventricles; this contraction starts at the base of the ventricles and pumps blood out of the heart at high pressure.

Diastole is when the ventricle muscle is **relaxed** and the atria are pumping blood into the ventricles. Systole is when the ventricles **contract**.

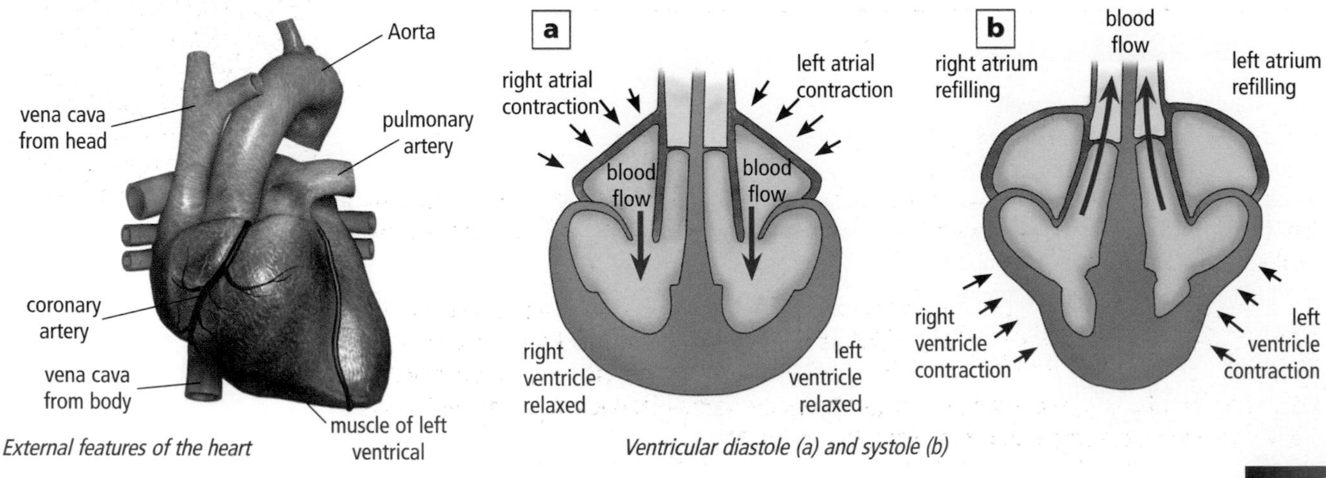

External features of the heart

Ventricular diastole (a) and systole (b)

contd

MEASURING FEATURES OF THE CVS

The CVS has some parameters that can be measured. These are defined in the table, which also shows the normal values for a healthy 20 year-old at rest.

Feature of CVS	Definition	Normal resting values
heart rate (HR)	the number of heart contractions in one minute	72 per minute
stroke volume (SV)	the volume of blood pumped out by one ventricle during one systole	0·07 litres
cardiac output (CO)	the volume of blood pumped by one ventricle in one minute (CO = HR × SV)	72 × 0·07 = 5 litres min^{-1}
blood pressure (BP)	the pressure during systole and diastole (measured in millimetres of mercury, mm Hg)	120 (systolic) 80 (diastolic)

DON'T FORGET

Normal blood pressure is described as '120 over 80' but it is not a fraction! The numbers are shown one above the other as a way of presenting both values together; the upper is the systolic pressure and the lower is the diastolic pressure.

EFFECT OF EXERCISE ON THE CVS

During exercise, the CVS has to **deliver more oxygen** to provide for increased respiration in the skeletal muscles. A variety of changes happen in the CVS to keep up with the demand for oxygen.

- **Cardiac output** (**CO**) increases.
- The **heart rate** (**HR**) increases.
- As HR increases, the atria receive more blood back from the body, so the atrial walls are more distended (stretched). This distension causes the force of the atrial contraction to increase, so more blood is pumped into the ventricles.
- The increased volume of blood delivered by the atria causes an increase in the distension of the ventricles, so the **stroke volume** of the ventricles is increased.
- The increased distension of the ventricles increases their force of contraction, so the systolic blood pressure increases.
- **Blood flow** to the working skeletal muscles is increased due to vasodilation.
- There is increased blood flow through the coronary arteries to keep the heart muscle aerobic.
- The blood flow to the gut and kidneys is decreased by vasoconstriction.
- And finally, since the body heats up during the exercise, the blood flow to the skin is increased by vasodilation. This increases the radiation of heat from the body's surface.

Blood flow to	Blood flow at rest (ml min^{-1})	Blood flow during exercise (ml min^{-1})
heart	250	1000
active skeletal muscles	650	20 850
brain	750	750
kidneys, liver and intestine	3100	600

Average blood flow for a man at rest and during exercise

The period of time it takes for the CVS parameters to return to normal after exercise is called the **recovery time**. During this period, the by-products of exercise are removed from the muscles and enter the bloodstream for transport to the kidneys, liver or lungs for processing or removal.

Search for 'measuring blood pressure' on YouTube for a 3.26 minute clip which will help you understand some of the ideas on this page.

LET'S THINK ABOUT THIS

The elastic nature of the artery walls does more than simply resist pressure. As the vessels swell with a pulse of blood following ventricular contraction, the walls recoil and convert the pulse into a continuous flow.

You should not go swimming immediately after eating a meal. Why is this advice important? After a meal, the digestive system is working hard and blood is directed there, away from the skeletal muscles. When you start swimming, your skeletal muscles work hard but the blood flow is not sufficient and the cells quickly run out of oxygen. Muscle fatigue and cramp set in – and when you are in the water and stop swimming, you sink!

TRAINING AND TESTING

EFFECT OF TRAINING ON THE CVS

Small bouts of exercise cause short-term changes in the cardiovascular system (CVS). These changes ensure that the respiring skeletal muscle cells are supplied with oxygen for aerobic respiration.

Physical training and repeated exercise cause **long-term changes** to the CVS. These changes are **fundamental adaptations** that allow the body to cope with the increased workload imposed by the training regime.

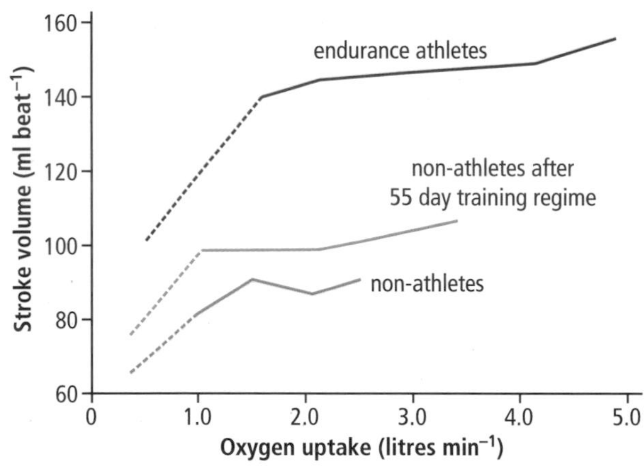

The effect of training on stroke volume. (Oxygen uptake is a measure of workload.)

Long-term changes to the CVS

During the course of an exercise training programme, the muscle mass of the heart increases. This increase in size is called **cardiac hypertrophy** ('heart over feeding'). The **left ventricle** shows the greatest increase as it is the chamber that supplies the increased demand from the muscles.

In endurance athletes, the left ventricle increases from 200 g to 300 g of muscle. In the cardiac muscle, the **muscle fibres become thicker** because each cell has **more contractile elements**, thereby increasing the strength of the cardiac muscle. Because the cardiac muscle is stronger, this leads to a **larger stroke volume** each time the left ventricle contracts. The data on the white graph show the effect of training on SV.

Since cardiac output (CO) is a function of heart rate and stroke volume (CO = HR × SV), the increase in SV creates a **higher maximum cardiac output**. Untrained individuals have a maximum CO of about 22 litres min^{-1}, while endurance athletes can reach a CO of 35 litres min^{-1}, an increase of about 60%.

Exercise training also increases the potential **maximal heart rate** during exercise. This is possible because the cardiac muscle is stronger. However, a more valuable feature of this increased strength is that the **heart rate remains lower** when the workload is increased. The data on the yellow graph supports the latter conclusion.

As you can see from the yellow graph, training also **reduces the resting heart rate**. For the non-athletes, the resting HR fell from 84 to 58 (a 31% decrease) after only 55 days training.

DON'T FORGET

The effect of exercise on the CVS is short term (see page 99). The effect of training is long term.

Training causes other physical changes that increase **oxygen delivery** and so maintain aerobic efficiency. In the heart, the myocardial circulation is improved by increasing the **density of the capillary networks.**

In the skeletal muscles that are being worked during the training programme denser capillary networks also develop, meaning that the muscles are less likely to become anaerobic. In a trained person, there is less build up of lactic acid during exercise, so the **recovery time** is much shorter than in an untrained person.

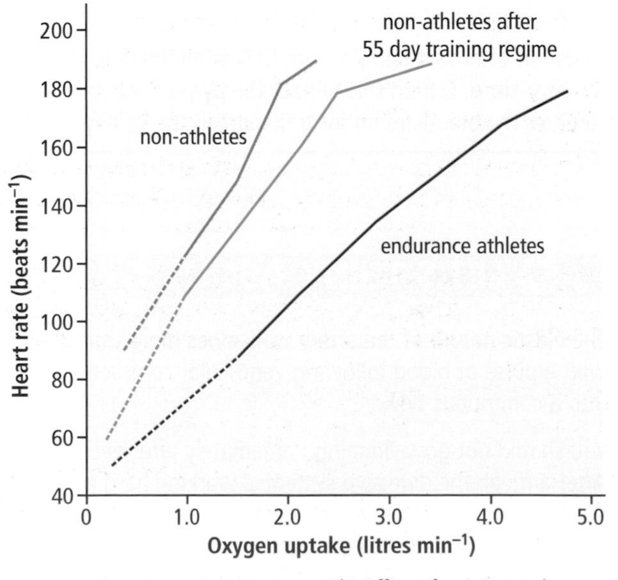

The effect of training on heart rate. (Oxygen uptake is a measure of workload.)

DON'T FORGET

Training causes adaptations to increase the supply of oxygen to respiring cells.

EXERCISE TESTING

The rate of oxygen uptake by muscles is the limiting factor in their performance. People with greater aerobic fitness have higher oxygen uptake into their muscles, so are able to work at higher intensities before fatigue sets in.

Exercise testing measures aerobic fitness by finding (or estimating) **$VO_{2\ max}$** – this is the maximum volume of oxygen used by the person, per kilogram of body mass, per minute. The inclusion of body mass in the units for $VO_{2\ max}$ makes the values comparable for different people. A higher value for $VO_{2\ max}$ indicates greater aerobic fitness. The two categories of exercise tests that are used are **maximal tests** and **sub-maximal tests**.

DON'T FORGET

$VO_{2\ max}$ is measured in ***$ml\,kg^{-1}\,min^{-1}$****.*

Maximal tests

These tests are used to monitor the effectiveness of training programmes which aim to improve an athlete's $VO_{2\ max}$. As the name suggests, maximal tests work an individual to **exhaustion**, while measuring their oxygen use. This makes these tests an accurate method of finding $VO_{2\ max}$. Because of the health risks of working to exhaustion, maximal tests require carefully controlled conditions and are only used under **medical supervision**. The test is usually carried out on a **treadmill** or **bicycle ergometer**. The athlete's oxygen use is measured as the workload on the machine is gradually increased until the oxygen use levels off at a maximum – this is the $VO_{2\ max}$.

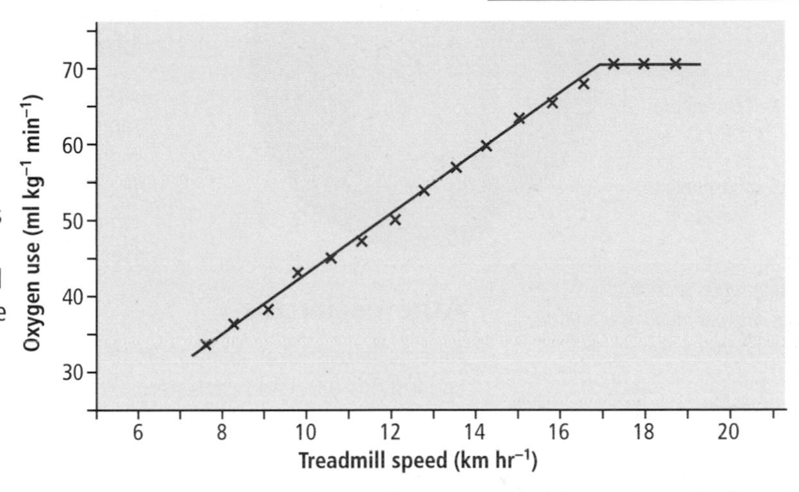

Oxygen use results for an athlete on a treadmill. ($VO_{2\ max}$ is 72 $ml\,kg^{-1}\,min^{-1}$.)

Sub-maximal tests

Sub-maximal exercise tests do not take the person to exhaustion, hence their name. They use **lower exercise levels** and then **extrapolate** the data to find an estimate for $VO_{2\ max}$. Clearly, this makes sub-maximal tests **less accurate,** but safer. For this reason, they can be used to monitor fitness of non-athletes or to monitor the recovery programme for patients with CVD (see page 103).

A good estimate can be obtained for an individual if their oxygen use and heart rate are measured while increasing the workload on a **treadmill** or **bicycle ergometer**. The exercise is stopped before exhaustion and the data is plotted on a graph (like the green graph). The **correlation** of heart rate with oxygen use can be used to **estimate** $VO_{2\ max}$ by extending the line to the estimated maximum heart rate, usually estimated as **220 minus the person's age in years**.

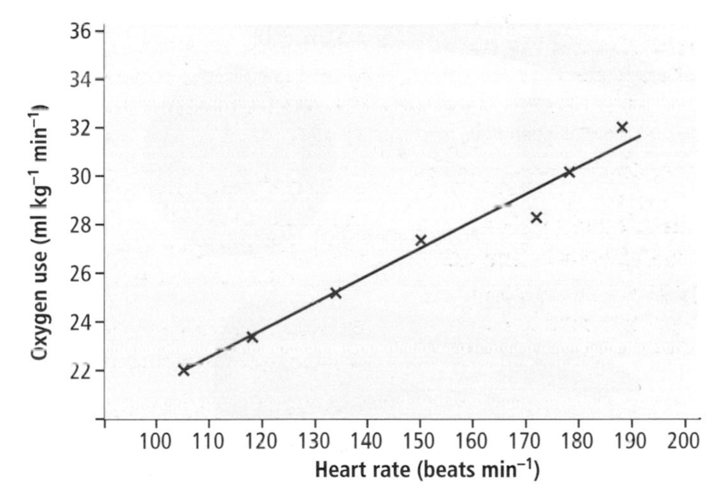

What is the estimated $VO_{2\ max}$ for this 20-year-old person?

Other sub-maximal tests include **step tests** and **beep test shuttle runs.** The results obtained from these are compared with values obtained from population surveys to give an estimated $VO_{2\ max}$.

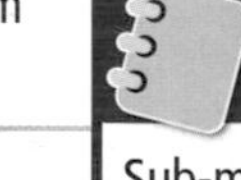

DON'T FORGET

Sub-maximal tests extrapolate data so they can only provide an estimate of the $VO_{2\ max}$.

Various researchers have suggested different methods of estimating maximum heart rate. Have yours estimated at http://www.brianmac.co.uk/maxhr.htm

LET'S THINK ABOUT THIS

A person's potential maximum heart rate increases as they get fitter. However, the yellow graph does not provide clear evidence to support this statement as the endurance athletes did not work to their maximum heart rate.

The estimated maximum heart rate is taken as 220 minus age in years, but this is only true for untrained individuals. In people who have done regular aerobic exercise, the maximal heart rate does not fall as quickly as they age.

CARDIOVASCULAR DISEASE (CVD)

PATHOPHYSIOLOGY OF CVD

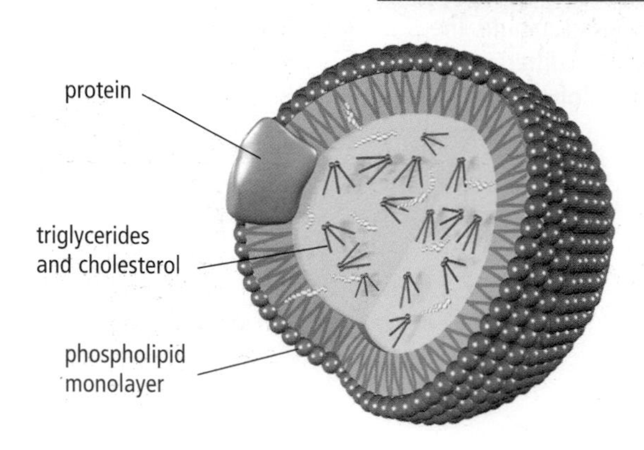

A lipoprotein showing the phospholipid monolayer and the hydrophobic pocket inside

Pathophysiology describes how diseases develop in the body. Cardiovascular disease (CVD) is a group of diseases of the heart and blood vessels. The main disease process that leads to CVD is **atherosclerosis**.

Lipoproteins

Lipoproteins are necessary for the transport of hydrophobic lipid molecules in the aqueous bloodstream. They are small 'bags' made up of a monolayer of phospholipids with some proteins floating in the layer. Within the bag is a hydrophobic pocket which can carry triglycerides and steroids such as cholesterol (see pages 14–15).

Atherosclerosis

Atherosclerosis takes many years to develop. If the wall of an artery is injured, low density lipoproteins (LDL) add cholesterol to the cells **under the endothelial lining** of the artery (see page 98). **Phagocytic cells** then move under the lining to remove the cholesterol. However, these phagocytic cells turn into **foam cells** which accumulate. Next, fibroblast cells move under the lining and deposit a fibrous material over the foam cells, so **fibrous material** builds up within the artery wall. This fibrous material gradually hardens due to a **build up of calcium**, leading to the formation of a hard **plaque** under the endothelial lining. As the plaque continues to build up, the lumen of the artery becomes narrower, thereby restricting the flow of blood.

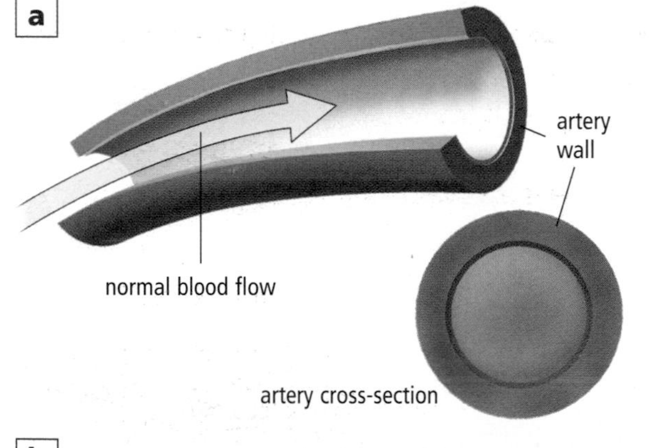

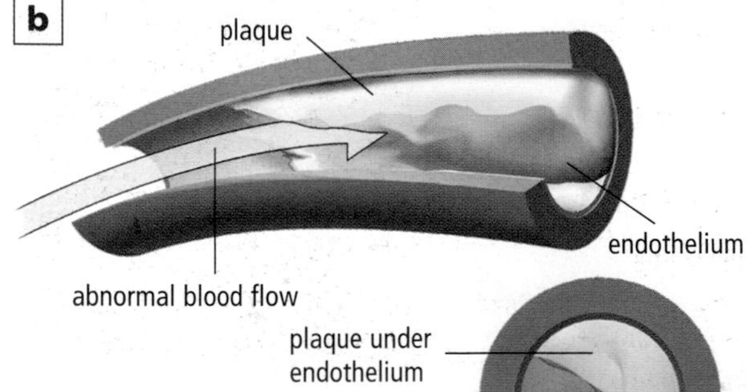

A normal artery (a) and one with atherosclerosis (b)

Hypertension

Hypertension is a condition in which blood pressure is persistently high. When blood pressure at rest is greater than $\frac{140}{90}$, hypertension would be diagnosed. Remember, $\frac{120}{80}$ is normal. Hypertension is a recognised cause of atherosclerosis as long periods of high blood pressure can cause the injury to the artery wall that allows the atherosclerosis to start.

Angina pectoris

When atherosclerosis happens in the coronary arteries, the narrowing due to the plaque restricts the blood flow to the **myocardium** (the heart muscle). When the heart is working hard, the decrease in blood flow to the myocardium may be severe and the cardiac muscle cells do not receive sufficient oxygen to maintain aerobic respiration. This causes anaerobic respiration, which is felt as **angina pectoris**, an extreme chest pain that fades again at rest as the heart rate slows and the cells receive sufficient oxygen.

Thrombosis and embolism

An atherosclerosis plaque makes the inner surface of the artery rough. **Platelets** can attach to this rough surface and then start to release clotting factors. This leads to the formation of a **thrombus** (blood clot) in the artery. If the thrombus blocks the artery completely, a **thrombosis** occurs, cutting off the blood supply to the cells that are usually supplied by that artery.

As the thrombus starts to build up, the increased blood pressure may break off a part of the clot. The free-floating clot is called an **embolus** and it is moved along the artery until it blocks a smaller part of the system. This blockage is called an **embolism**.

DON'T FORGET

A thrombosis is caused by a thrombus attached to the plaque.

contd

PATHOPHYSIOLOGY OF CVD contd

Myocardial and cerebral infarctions

Cardiac muscle cells and brain cells use oxygen very quickly, so their function is impaired when oxygen delivery is interrupted. **Myocardial infarction (heart attack)** is caused by a thrombosis or embolism in the coronary artery. The area of cardiac muscle that receives blood from this artery dies due to lack of oxygen. A **cerebral infarction (stroke)** can be caused by a thrombosis or embolism in a brain artery; the brain tissue also dies due to lack of oxygen.

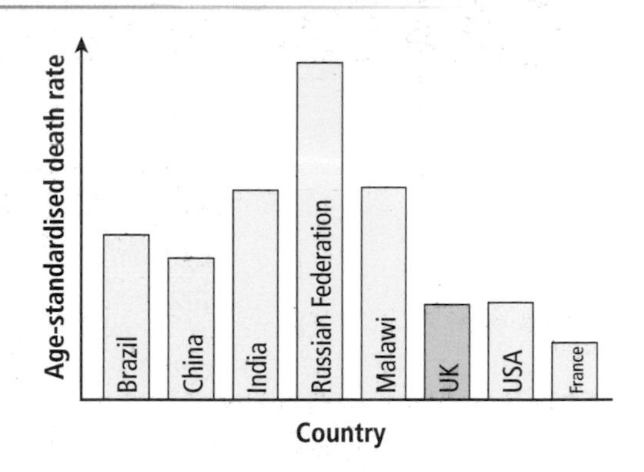

Deaths due to CVD in 2002 in a selection of countries. The data from the World Health Organisation shows ***deaths*** *due to CVD, not the* ***incidence*** *of CVD. Is better health care causing the low values for USA, UK and France?*

Incidence of CVD in UK and other countries

In recent years the incidence of deaths due to CVD in the UK has fallen. In 1994, there were about 500 deaths per 100,000 of the population due to CVD, whereas the figure for 2002 (shown in the graph) has fallen to about 200 per 100,000. This still makes CVD one of the largest killers in the UK, causing about a third of all deaths of people under 75 years old.

CVD risk factors

Non-modifiable risk factors (physiological)	
Factor	**Increased risk**
age	atherosclerosis increases for males over 45 years old and females over 55 years old
gender	men accumulate more LDL in the blood
heredity	family history of CVD due to inherited genes
race	African or Indian ancestry

Modifiable risk factors (lifestyle)		
Factor	**Increased risk**	**Reduce risk**
diet	high fat diet causes more LDL in blood	reduce fat intake
smoking	carbon monoxide makes platelets more likely to cause clots	stop smoking
obesity	more LDL in circulation	more activity AND less energy intake
inactivity	heart muscle is less adapted for effort	regular exercise

BENEFITS OF EXERCISE

A regular programme of exercise has **protective effects** against the development of CVD. Exercise is also used in the **rehabilitation of CVD patients**.

Regular exercise produces the beneficial effects of physical training (see page 100), improving the efficiency of the heart in two ways:

1. The myocardial circulation is improved by development of denser capillary networks, **protecting the heart against oxygen deficiency**.
2. The contractile properties of the myocardium are enhanced by increasing the thickness of the muscle fibres. This results in a lower heart rate at rest and during exercise.

Regular exercise uses more energy and so helps to **reduce the percentage body fat**, also reducing the chance of obesity, a risk factor for CVD. Exercise also causes an increase in high density lipoproteins (HDL) in the blood – HDL reduce the blood cholesterol level and, therefore, the risk of CVD. With rising HDL, there is a corresponding decrease in low density lipoproteins in the blood (LDL). LDL are involved in the development of atherosclerosis, so a decrease in LDL also reduces the risk of CVD.

Search in YouTube for 'artery explorer'. This 5.06 minute animation covers most of the points on this page brilliantly (although it does talk about good and bad cholesterol!).

DON'T FORGET

It may be useful to remember HDL as 'helpful', and LDL as 'lethal'. However, you should not refer to them as 'good cholesterol' and 'bad cholesterol'!

LET'S THINK ABOUT THIS

Another risk factor for CVD is hypertension, but is it modifiable or non-modifiable? A high salt intake, smoking and obesity all increase the chances of developing hypertension, and these can all be modified. But about 70% of hypertension cases report other affected family members suggesting an inherited aspect that cannot be modified.

ENERGY AND METABOLISM

ENERGY BALANCE

The main components in our diet are carbohydrates, lipids and proteins. These contain **chemical energy** that can be released during cellular respiration to **synthesise ATP**. The energy in food is measured in **kilojoules per gram** ($kJ\,g^{-1}$).

Component of diet	Energy content
carbohydrate	$17\,kJ\,g^{-1}$
fat	$38\,kJ\,g^{-1}$
protein	$17\,kJ\,g^{-1}$ (average value)

The energy present in different components of the diet. (Protein has an average value because proteins vary in the number and type of amino acid present.)

The energy released from food is used for a range of functions in the body. These include: muscle contraction, synthesis of protein, maintenance of body temperature, active transport, and transmission of nerve impulses.

In ideal conditions, the human body balances the food energy taken in with its energy output. However, this balance is not always maintained and any imbalance causes a change in the body's energy stores. The effect of an imbalance can be represented as:

energy IN – energy OUT = change in energy stores

If less energy is taken in than expended, the body uses up fat from its energy stores. If more energy is taken in than the output, fat is built up in energy stores.

Most of the fat molecules stored in the human body are saturated triglycerides (see page 14).

A high energy diet and lack of activity cause excess fat to be stored in the body, potentially leading to **obesity** (see page 107), a risk factor for cardiovascular disease (CVD). Obesity is linked to an increase in low density lipoproteins (LDL) in the blood; LDL have a major role in the process of atherosclerosis which leads to CVD (see page 102). Dietary recommendations for health always recommend **reducing the fat content** of the diet. Switching to a healthy diet has been estimated to reduce the risk of developing CVD by about 30%.

BASAL METABOLIC RATE (BMR)

The basal metabolic rate (BMR) is the total **energy expended for basic body functions**. The BMR makes up about **60–70% of the total daily energy output** in a sedentary adult. The average BMR for adult females is about $6000\,kJ\,day^{-1}$ while for adult males it is about $7000\,kJ\,day^{-1}$. There is, however, considerable variability in BMR between individuals as shown on the graph.

BMR ($kJ\,day^{-1}$): 10000, 8000, 6000, 4000, 2000, 0
Body mass (kg): 0, 50, 100
males (n = 2821)
females (n = 1664)

Statistically-produced lines of best fit for the effects of body weight and gender on daily BMR for 18–30 year-olds. Note the huge sample size for each line.

basal metabolic rate (breathing, contraction of heart muscle, production of body heat, kidney function, cell metabolism)

additional energy expenditure (physical activity, thermoregulation, digestion of food)

*Total energy expenditure is **BMR** and **additional energy expenditure***

Measuring BMR

All the energy used in the body is ultimately lost as heat, so heat release (in $\mathbf{kJ\,day^{-1}}$) can be used to measure basal metabolic rate. BMR can also be measured as energy use per kilogram of body mass per day ($kJ\,kg^{-1}\,day^{-1}$) to allow comparison between individuals.

BMR is measured under standard conditions which remove the additional energy uses. The person being tested should be: (1) completely rested, both before and during the measurements; (2) fasted for at least 10–12 hours before the measurements are taken; (3) in a thermo-neutral environment (22–26°C) so the person does not need to thermoregulate.

Find out more about the BMR at http://www.answers.com/topic/basal-metabolic-rate.

contd

BASAL METABOLIC RATE (BMR) contd

Factors affecting BMR

Even taking account of these factors during the measurement of BMR, the values obtained for two people can be widely different, even though the people may appear similar. This is because BMR is affected by a range of factors.

- **Body mass** – as mass increases, BMR increases because there are more actively respiring cells in a larger person.
- **Body composition** – more lean tissue and less fat per kg of body weight leads to a higher BMR because lean muscle tissue is more metabolically active than fat tissue per gram of mass.
- **Age** – BMR is higher in children (who use energy for growth) and, after age 20, it decreases by about 2% every 10 years.
- **Gender** – BMR is higher in males, on average, because males tend to have more lean tissue and less fat per kg of body weight than females.
- **Nutritional status** – a diet with a low energy intake (such as during fasting or dieting) will cause a decrease in BMR.

ADDITIONAL ENERGY EXPENDITURE ABOVE BMR

Total daily energy expenditure includes the BMR plus additional energy expenditure.

- **Physical activity** above resting increases additional energy expenditure and, as intensity and duration of the activity increase, so does energy expenditure.
- A **colder climate** increases additional energy expenditure. As temperature decreases, heat loss increases. The body increases its energy expenditure to generate more heat to maintain body temperature.
- **Dietary-induced thermogenesis** is the additional energy expenditure needed to digest food. This makes up about 10% of total daily energy expenditure, depending on the food types eaten.
- The **composition of the diet** also affects energy expenditure. High protein diets and high carbohydrate diets have high energy expenditure for digestion. High fat diets have low energy expenditure.
- During **pregnancy**, there will be energy requirements for the growth of the foetus, placenta and breast tissue. Similarly, **lactation** requires energy for the synthesis of molecules for the milk.

MEASURING ENERGY EXPENDITURE

All the energy expenditure in the body is ultimately released as heat. **Direct calorimetry** measures the **heat released** from the body in an airtight chamber, which is insulated to prevent heat loss. A person lives in the chamber for several days and the heat released from their body is found by comparing the temperatures of water entering and exiting a **heat exchanger**.

There is a direct correlation between oxygen consumption and energy expenditure, so **indirect calorimetry** can be used to measure energy expenditure. Here **oxygen consumption** is measured over a period of time. The total volumes of oxygen inhaled and exhaled are compared. To do this, the total volume and the % oxygen content are measured for both inhaled and exhaled air, so that the total volume of oxygen taken in can be calculated. The exhaled air can be collected in a large bag for analysis, or a portable respirometer can be used to measure oxygen consumption electronically.

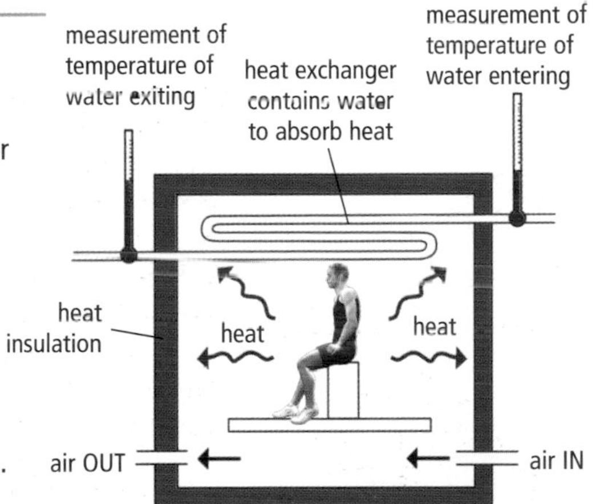

Direct calorimetry

The most indirect measurement of energy expenditure is by **recording heart rate**. This is based on the correlation between heart rate and oxygen consumption (see the green graph on page 101). For this, the person's oxygen consumption and heart rate have to be found at various exercise workloads and, even then, the energy expenditure is estimated from two correlations.

DON'T FORGET

Direct calorimetry measures energy expenditure by measuring the heat loss directly. Indirect calorimetry estimates energy expenditure using a correlation to another form of energy measurement.

LET'S THINK ABOUT THIS

Pregnancy would be expected to increase the total energy expenditure. The estimated additional energy requirement over the whole pregnancy is estimated at about 350 000 kilojoules but the woman's total energy intake may not increase by anywhere near as much as this, if at all! Since her BMR (kJ kg^{-1} day^{-1}) remains at pre-pregnancy levels, the additional energy requirement is probably made up through a reduction in activity.

BODY COMPOSITION AND WEIGHT CONTROL

MEASURING BODY COMPOSITION

Body composition measurements are used to find the proportion of fat to lean muscle tissue in a person. Some techniques are accurate but difficult to perform, while others are easier to perform but are based on estimates and correlations.

Densitometry

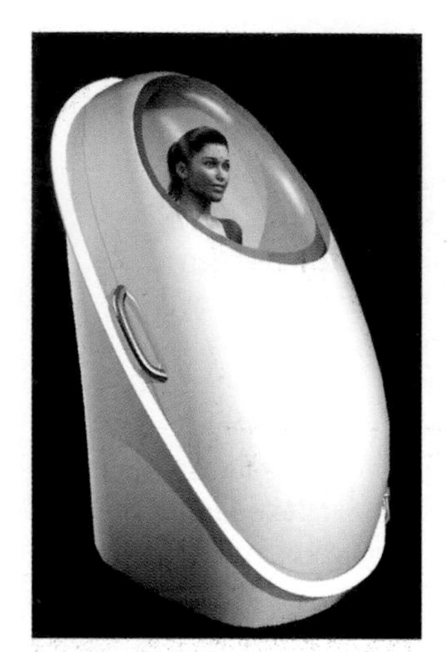

A Bodpod being used to measure body volume

Density is the mass (g) divided by volume (cm^3). Fat has a lower density than lean muscle tissue (that is 1 cm^3 of fat weighs less than 1 cm^3 of muscle), so a higher percentage of body fat leads to a lower body density. To find a person's density, their body mass and volume must be measured. The density is then used in the **Siri equation** to find the percentage body fat.

$$\text{percentage body fat} = \left(\frac{495}{\text{density (g cm}^{-3})}\right) - 450$$

Weight can be found easily using a set of scales, but volume is a more difficult measurement. One way to find the volume is by using a **BODPOD**, a sophisticated piece of equipment that measures air displacement caused by a body.

Underwater weighing is simpler and uses water displacement to find the body's volume. The person is weighed on land and then is weighed as they are suspended underwater in a seat. The body volume is the difference between the land weight and the immersed weight, since 1g of buoyancy is equal to 1 cm^3 of body volume.

Densitometry is the most accurate method but it has limitations: a BODPOD is very expensive (more than $30 000 in 2010) and underwater weighing is a difficult procedure as some people may be unwilling to be completely immersed underwater.

Skinfold thickness

Skinfold callipers are used to measure the thickness of a skinfold (two layers of skin with the fat between them) at four different skinfold sites. These values are then used in a table (which takes account of age and sex) to give the person's percentage body fat. However, skinfold thickness measurements are difficult to take accurately. Skinfold thickness also does not take account of unusual fat distribution and is particularly difficult to carry out on very obese or very lean individuals.

Bioelectrical impedance analysis (BIA)

Bioelectrical impedance is the electrical resistance of the body to a small applied electrical current. The current travels through the water in the body and, since fat cells contain much less water than other cells, fat acts as an electrical insulator. As the proportion of body fat increases, the impedance to the electrical current also increases.

The accuracy of BIA is affected by the hydration level of the person. Also BIA tends to over-estimate the fat content in lean or muscular people and can under-estimate the fat content in obese people.

DON'T FORGET

Each of the methods of measuring body composition has limitations.

Simple measurements

Body mass index (BMI) is found by dividing body mass (kg) by height squared (m^2). The categories defined by BMI ranges are shown in the table. BMI doesn't measure body composition, it just indicates whether an individual is relatively under or overweight. Care has to be taken with BMI measures as they tend to classify as overweight, people who actually have a very low percentage body fat. Athletes have proportionally more muscle tissue than normal, so will be heavier than others relative to their height.

BMI range	Category
less than 18·4	underweight
18·5 – 24·9	normal
25·1 – 29·9	overweight
30·0 – 39·9	obese
more than 40·0	severely obese

contd

MEASURING BODY COMPOSITION contd

The **waist:hip ratio** gives an indication of the distribution of fat in the body and is found by dividing the waist circumference by the hip circumference. A higher waist:hip ratio indicates that more fat is stored around the body's internal organs and this has been found to be strong indicator of an increased risk of heart disease. A greater health risk is associated with a waist:hip ratio of >0·95 for men or >0·8 for women. However, there can be errors in the measurements if the exact same position is not measured for each person.

Measuring the **mid-upper arm circumference** gives a value which can be compared to standard values to give a rough estimate of body fat content. This method is often used in famine relief to help make quick decisions about which people need treatment first.

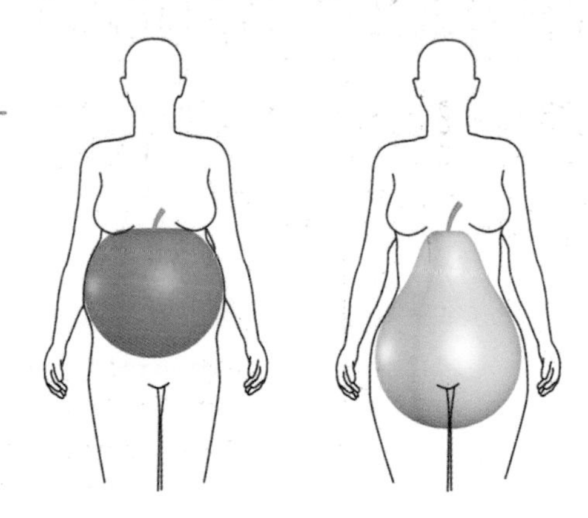

A high waist:hip ratio gives an 'apple-shaped' body as there is extra fat around the waist. A 'pear-shaped' body has extra fat around hips and thighs and a low waist:hip ratio.

OBESITY

Obesity is a severe excess of body fat and is characterised by a **BMI of 30 and above.** The incidence of obesity in the UK has been steadily rising. More people are storing body fat because energy intake has increased due to an **energy-dense diet** (high in fats and sugars) and energy output has decreased due to a **decrease in physical activity** (see page 104).

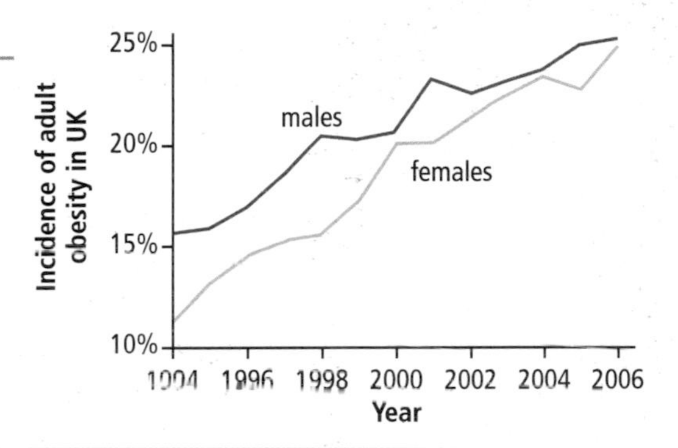

Incidence of adult obesity in UK, 1994–2006

EXERCISE AND WEIGHT CONTROL

An exercise regime increases the energy output relative to energy input, increasing the loss of fat from the body's energy stores (see page 104). The **percentage body fat decreases**, but the impact diminishes as the exercise regime continues (as shown on the graph on the right). The exercise regime also **preserves or increases lean muscle tissue**. If the lean muscle tissue increases, the basal metabolic rate (BMR) also increases because muscle tissue is more metabolically active than fat tissue.

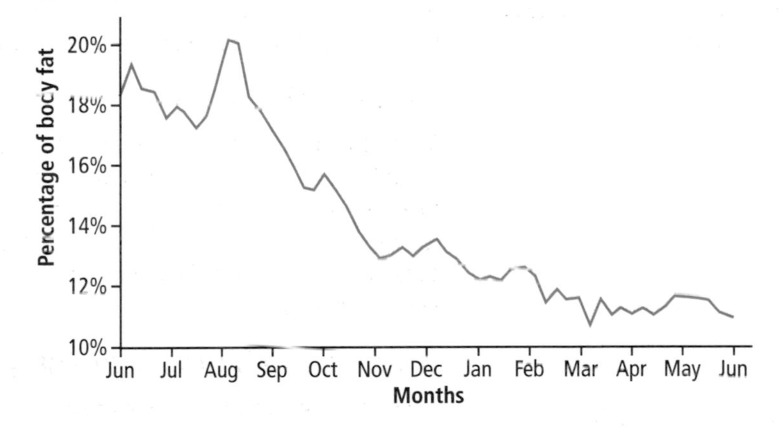

Changes in an individual's percentage body fat during a year's training

The most effective type of exercise for weight loss is **aerobic exercise** (such as brisk walking, jogging, swimming or cycling) at a **moderate intensity**. To be most effective, the exercise regime should comprise **frequent** exercise (four or five times a week) for a **long duration** (30 minutes or more). However, depending on an individual's initial fitness level, it may be necessary to build up frequency, intensity and duration gradually.

Go to http://www.rowett.ac.uk/edu_web/sec_pup/body_comp.pdf to find out more about measuring body composition. You could also watch a BODPOD in action at http://www.youtube.com/watch?v=elobnbT33yo

DON'T FORGET

Exercise for weight loss needs to be **famil-d** (**f**requent, **a**erobic, **m**oderate **i**ntensity, and **l**ong **d**uration).

LET'S THINK ABOUT THIS

It has been suggested that BMR increases for a few days after a bout of exercise, but the evidence for this is still inconclusive. However, it is clear that BMR is raised for a few days after extremely strenuous exercise, possibly for repair of minor injuries to the muscle tissue.

OSTEOPOROSIS AND DIABETES

OSTEOPOROSIS

Normal bone growth

Children need a diet with sufficient Vitamin D and calcium to help build bones. As more **calcium is deposited**, the bone becomes denser and stronger. **Bone density** continues to increase from late adolescence and peaks in the mid-20s.

Bone tissue is a living material which is constantly being **remodelled** by osteoclast cells (which remove old bone material) and osteoblast cells (which reconstruct new bone material). The level of osteoclast activity is regulated by oestrogen levels in females and males. In a young adult, all the bone cells in the skeleton are replaced in about seven years.

Development of osteoporosis

From age 30 onwards, the rate of bone material replacement by osteoblasts gradually decreases, leading to a **progressive loss of calcium** from the bones. As a result of this loss of calcium bone density gradually decreases.

Osteoporosis occurs when **loss of bone density** causes bones to become more **porous** and **brittle**, and hence more **susceptible to fracture**. Common symptoms include loss of height and curvature of the spine, caused by many small fractures in the vertebrae. In the UK, osteoporotic fractures of the wrist, hip or spine occur in about 50% of women over 50 and about 20% of men over 50. Osteoporosis can also affect children but this is rare.

The rate of loss of bone density is much slower in men than in women, so osteoporosis occurs earlier in women. It is most common in **post-menopausal** women. After the menopause, the production of oestrogen decreases. As a result, osteoblasts become less active and loss of bone density increases significantly.

Effect of exercise on the development of osteoporosis

Osteoporosis is due to a gradual loss of bone density.

Physically fit people have **greater bone density** and so have a lower risk of osteoporosis. Exercise puts the bones under a degree of stress and this increases calcium deposition by osteoblasts, thus increasing bone density. The best exercise for prevention of osteoporosis is **regular weight-bearing exercise** of a **moderate intensity**, such as running, walking, or dancing.

Women are recommended to use exercise to maximise their bone density in their 20s and 30s. This **reduces the risk** of osteoporosis in later life, as bone density is promoted before age-related loss starts to have an effect. Exercise is also recommended for post-memopausal women as it helps to **maintain bone density** and so **delays the progress** of osteoporosis.

The effects of exercise are not always beneficial. Young female endurance athletes may develop osteoporosis if their oestrogen level decreases, causing osteoblasts to become less active.

DIABETES

Control of blood glucose

An **in**crease in blood glucose levels causes **in**sulin secretion to be **in**creased so that more glucose is taken **in** to the liver cells.

Blood glucose level must be kept within narrow limits, between 3·9–6·1 mmol per litre of blood. The hormones **insulin** and **glucagon** interact in a negative feedback system to control blood glucose levels. Both these hormones are peptide hormones, so they are hydrophilic signalling molecules which bind to receptors on their target cells (see page 31). Insulin and glucagon receptors are found in the membranes of liver cells, muscle cells and others.

After a meal, blood glucose levels increase. The pancreas detects the **increase in blood glucose levels** and this causes both an increase in insulin secretion and a decrease in glucagon secretion by the islets of Langerhans in the pancreas. Insulin lowers the blood glucose level by binding to receptors in liver cells. This has two effects: first, it stimulates the **uptake of glucose** into liver

contd

DIABETES contd

cells; second, it stimulates **glycogen synthesis** in liver cells, so excess glucose is converted to glycogen (see page 13).

Gradually the blood glucose level decreases as the liver cells take in glucose and the muscle cells use glucose for respiration. This **decrease in blood glucose level** is detected by the pancreas and causes increased glucagon secretion and decreased insulin secretion. Glucagon binds to receptors in the liver cells which stimulate the **degradation of glycogen** to glucose, and so more glucose is released into the bloodstream; blood glucose levels are maintained.

pancreas secretes more insulin
pancreas secretes less glucagon
insulin increases
- glucose uptake by liver
- synthesis of glycogen

detected by pancreas
increasing blood glucose levels
decreasing blood glucose levels
glucagon increases degradation of glycogen
detected by pancreas
pancreas secretes more glucagon
pancreas secretes less insulin

Negative feedback control of blood glucose levels

Non-insulin dependent diabetes mellitus (NIDDM)

Diabetes mellitus is a medical condition caused by a deficiency in the effect of insulin, resulting in the loss of control of blood glucose level. Type 1 diabetes is insulin-dependent diabetes mellitus (IDDM), caused by a failure to produce insulin; treatment is with insulin injections.

Type 2 diabetes is **non-insulin dependent diabetes mellitus** (**NIDDM**) in which normal insulin levels exist in the blood plasma. However, the skeletal muscle cells have less sensitivity to insulin, and so they have a reduced uptake of glucose from the blood. This is most likely to be because the cells are deficient in active insulin receptors. This lack of sensitivity to insulin is called **insulin resistance**.

NIDDM used to be relatively rare and found only in people over 40 years old. In recent years, it has been increasing in prevalence in the UK and is being found in much younger people. This is because NIDDM is **strongly linked to obesity,** which is also becoming increasingly common; the majority of people diagnosed with NIDDM are overweight or obese individuals.

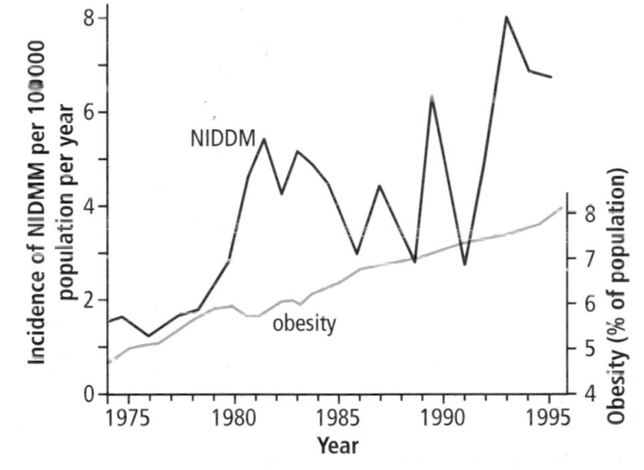

Annual incidence of NIDDM and prevalence of obesity in Japanese schoolchildren

Effect of exercise training on regulation of NIDDM

Exercise training can be used effectively in the treatment and regulation of NIDDM. Regular aerobic exercise can help to reduce obesity (see page 107) and, therefore, also reduces the risk of NIDDM.

Exercise training has three further effects which help to regulate NIDDM.

- It improves the uptake of glucose from blood into skeletal muscle cells by increasing the number of active insulin receptors in these cells.
- It causes an increase in the density of the capillary network in the skeletal muscles (see page 100); there is improved blood flow and so better glucose availability.
- It causes an increase in enzymes associated with glycogen synthesis, so glucose storage is enhanced.

A couple of clips on YouTube might help your understanding. Search for 'Osteoporosis 4' for a 3.09 clip on prevention of osteoporosis. Search for 'Dr Knope diabetes' for a 1.55 clip with physician's view on the treatment of NIDDM (Type 2) diabetes.

LET'S THINK ABOUT THIS

Osteoporosis and NIDDM seem to be influenced partly by multiple genetic components that increase susceptibility to these diseases. However, in both conditions major environmental components are also necessary for the disease to develop.

The risk of osteoporosis is increased by a calcium-deficient diet during childhood. The risk of developing NIDDM is high in Mexican Americans in the USA (due to high fat diets), but Mexicans who live in non-Westernised areas tend not to get NIDDM, no matter how high their genetic risk.

PLANNING AN INVESTIGATION

Gould's observation that Darwin's blackbirds, grosbeaks, finches and warblers from the Galapagos were, in fact, all closely-related finches has led to a very rich vein of scientific investigation and enquiry.

INVESTIGATIVE SCIENCE

Science consists of a body of knowledge that is modified through the careful evaluation of evidence. It is through investigation that evidence is gathered. New scientific ideas and details are added to the existing body of knowledge and previously accepted ideas are updated or superseded.

Ideas and thinking can change rapidly in science because there is a culture of sharing the findings of scientific research. Only by publishing full details of methods and results, can scientists hope to persuade others to share their conclusions. As this openness allows others to repeat the analysis, or the whole experiment, there is a tendency towards a high degree of honesty and integrity in the reporting of results. The evidence-based approach of science has placed it as the central philosophy of the modern world.

OBSERVATION LEADS TO INVESTIGATION

Even the simplest of observations can lead to very worthwhile investigations. Mendel's observations led to an understanding of the laws of inheritance. Gould's observations of Darwin's finches led to what is still one of the best-understood examples of natural selection, speciation and adaptive radiation.

Observations alone are not enough. Investigation involves the meticulous collection of data, under carefully controlled conditions, followed by impartial evaluation. Mendel grew 29 000 pea plants before publishing his results; it took Darwin 22 years of further collation of information after Gould's breakthrough to feel he had sufficient evidence to publish the *Origin of Species*.

DON'T FORGET

Ask your teacher to show you a copy of the *Suggestions for Investigations* document produced by SQA in 2005.

SEEKING INSPIRATION

It can be hard to select a suitable topic for investigation. Remember, the key point is that you are trying to answer a question, not just carrying out a practical with known results. Try to make sure that you are investigating a question which **interests** you, but also **listen** to the advice of others who have more experience of what is likely to be feasible. Also, your teachers may have particular skills and experience that you could use to your advantage. The internet will be invaluable during your early stages of planning. Look through some of the websites of the following organisations for advice and ideas:

- Science and Plants for Schools (www-saps.plantsci.cam.ac.uk)
- National Centre for Biotechnological Education (www.ncbe.reading.ac.uk)
- Microbiology in Schools Advisory Committee (www.microbiologyonline.org.uk/misac.html)
- Association for the Study of Animal Behaviour (asab.nottingham.ac.uk)
- Mystrica Practical Solutions (mystrica.com)
- British Ecological Society (www.britishecologicalsociety.org).

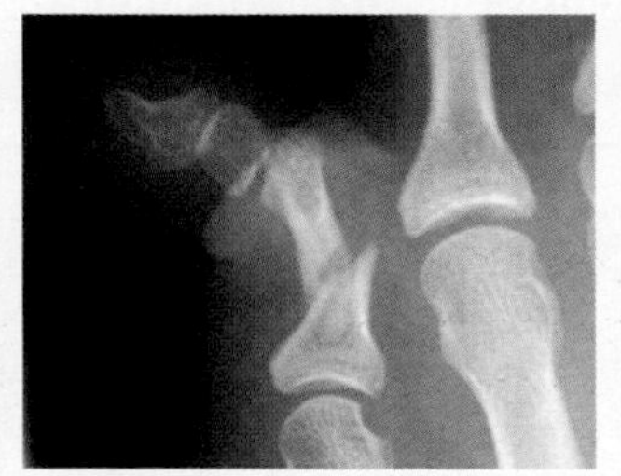
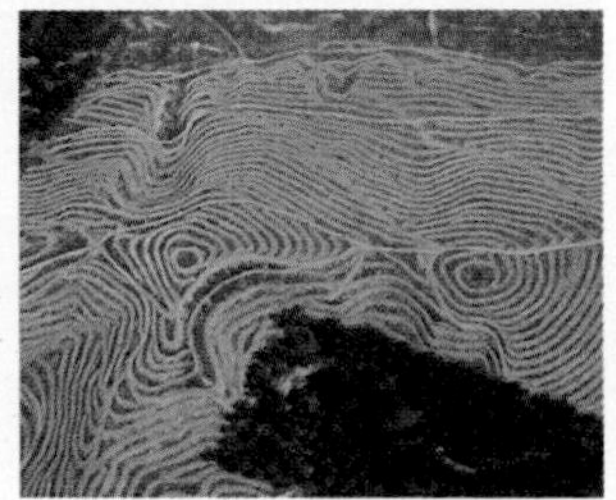

Inspiration for an investigation can come from any interesting observation. You could make a list of experiments or topics that you have enjoyed studying in the past. Maybe something sparked your interest on a fieldtrip? Or perhaps you have a special interest in a particular field of biology, such as medicine? Maybe you would like to investigate a topic that is related to the big challenges that society faces, such as habitat destruction and the control of the carbon cycle?

contd

EXPERIMENTAL DESIGN

Interesting observations raise interesting questions, and the purpose of scientific investigation is to provide evidence so that attempts can be made to answer these questions. To ensure that it is possible to draw conclusions from data, it is essential that due care and consideration is given to the following aspects of experimental design.

DON'T FORGET

Do not confuse a **control experiment** with the **control of confounding variables**.

Aim and hypotheses

While it is a good idea to trial experiments at an early stage in the planning, it is essential to have a very **clear aim** and hypothesis before the full experiment is begun. The aim is often expressed as 'to investigate the effect of x on y'. The hypothesis is a **prediction** and would be written something like 'as x increases, y is expected to decrease'.

Variables – things that can vary

In general, good experimental design involves changing one variable at a time. The variable that is changed (x) in the experiment is the **independent variable**. The variable that is measured is the **dependent variable**; these measurements are known as results (y). To work out whether the change in the independent variable has caused an effect in the dependent variable, it is essential that all other likely **confounding variables** are kept the same (**controlled**). For comparison, the independent variable can be removed entirely; this set-up is known as a **control**.

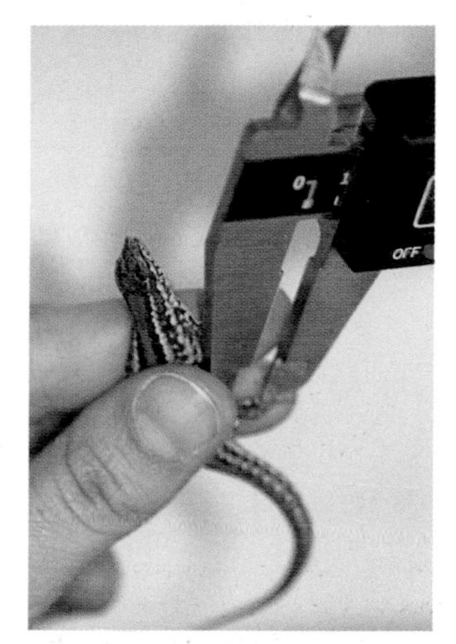

Think carefully about how you can make accurate measurements of your variables. In this study, rather than using a ruler, a digital vernier calliper (with an accuracy of 0.02 mm) was used to record lizard toe length. However, given the errors involved in measuring lizard toes, results were recorded and analysed to 0.1 mm accuracy only.

Accuracy – how close measurements are to reality

Accuracy of **measurement** is essential in experimentation. The different input values of the independent variable must be measured, as must any changes in the results of the dependent variable. In addition, it is also very important to measure the important confounding variables during the experiment to ensure that they are being controlled (that is, they are not varying). Investigators must ensure that their measurement techniques are sufficiently precise and lacking in bias to be able to distinguish differences in the values of their variables from any differences caused by human or instrument error. The accuracy of your measurements will clearly affect your ability to draw conclusions. Be aware that you will be penalised if your averaged results have more decimal places than your raw data!

Reliability – how typical are the results?

To keep an experiment manageable a **sample** is selected. However, a major source of variation in biological data is often the real differences between biological samples. To reduce the effect of either chance or natural variability skewing the interpretation of results, it is very important that the sample size is large enough so that any variation in the samples is representative of variation in the whole population. In addition, the whole experiment should be set up again and **repeated** (replicated) to check that the results are reliable (typical).

DON'T FORGET

In an investigation you are trying to answer a question.

Validity – is the experiment worth doing?

An experiment is valid if the choice of variables is appropriate, and also if the procedures chosen to alter, measure and control them are suitable. Many experiments are ruined by flawed techniques and poor choice of which variable to alter and how.

LET'S THINK ABOUT THIS

Only by thinking very clearly about variables can invalid experimental designs be avoided. It is very important that you **operationalise variables** before you start to collect data. This means that you decide exactly what is being altered and measured in your experiment. For example, in a seedling germination experiment, the dependent variable could be measured as height, biomass, dry mass, or even a particular measure of chlorophyll content.

DATA COLLECTION AND ANALYSIS

KEEPING A LABORATORY NOTEBOOK

All good scientists keep a clear record of everything they do, so that procedures or results are not forgotten. The NAB Assessment for the Investigation half-unit requires you to complete an **Investigation Notebook** as evidence that you have actually designed and carried out your own investigation.

Investigation NAB Outcome 1: Develop a plan for the investigation

- You must maintain a regular record.
- You must state the aims of your investigation clearly.
- You must formulate hypotheses or questions relevant to the aims of your investigation.
- You must use appropriate experimental, observational and sampling procedures, techniques and apparatus.
- You must consider the need for controls and replicate treatments or samples.
- You must consider any relevant problems associated with the use of living materials or natural habitats.

The benefits of using a laboratory notebook are greatest during the data collection and analysis phase of an investigation. All raw results are collated in one place, rather than on loose sheets. Any trends in the data can be explored by drawing sketch graphs in the laboratory notebook. Also, various different approaches to the analysis of the data can be trialled in the notebook and, if all calculated values are recorded there, you are less likely to have to repeat the calculations later.

The laboratory notebook is used as **evidence** that you have actually designed and carried out your own investigation. You must satisfy the nine different criteria listed on this and the previous page in order to pass this part of the course. Your laboratory notebook will be marked by your class teacher, and is the NAB for this half-unit. Your notebook may be checked by a verifier from SQA during the course and at the end.

NAB Outcome 2: Collect and analyse information obtained from the investigation

- You must collect experimental data with due accuracy.
- Relevant measurements and observations must be recorded in an appropriate format.
- Data must be analysed and presented in an appropriate format.

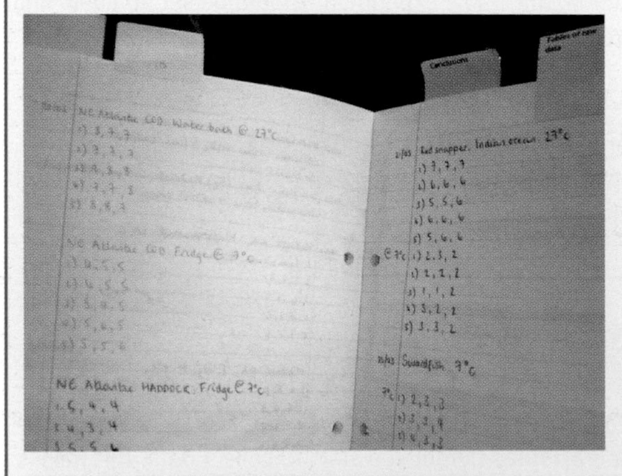

Get into the habit of recording your raw data in your laboratory notebook as soon as you measure it, no matter how memorable it seems at the time.

Get advice on your laboratory notebook by reading the Biology Internal Assessment Reports (Senior Verifier's Reports) for Advanced Higher Biology Investigation located on the SQA website.

contd

ANALYSING DATA

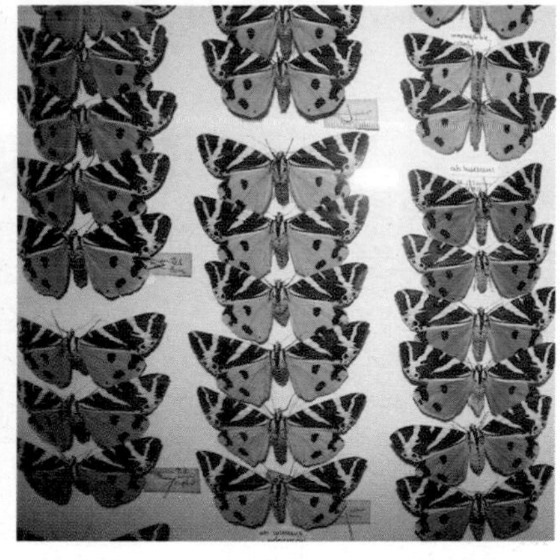

Data gathered in biology investigations normally includes much variation. This, of course, is due to the natural variability found in organisms.

Biological data contains much natural variation – organisms tend to be genetically variable, after all. The skill of designing and carrying out biological investigations is being able to distinguish between the natural variation in the sample and any differences in dependent variables that are the result of different treatments of the independent variable.

The usual method for dealing with variable data sets is to calculate the (mean) average. As well as this 'mid-point', at Advanced Higher level it is important to think about the variability of each sample data set. For this reason, ranges, standard deviations or standard errors are often calculated and plotted alongside the mean. Differences between samples or treatments are often confirmed, not only by differences in mean averages, but by a lack of overlap in error ranges.

Analysis top tips

When analysing your data, essentially you are asking one main question: 'Is my independent variable causing an effect in my dependent variable?' To answer this question, the variation within each sample is assessed. A narrow range of variation within a sample indicates that the average is reliable, especially if the sample size is large.

Differences between treatments are then compared. Large differences, without any overlap in the ranges of variability, indicate that the differences could be real effects – that is, the independent variable does influence the dependent variable. However, the replicate data set(s) also have to be analysed to double check that the effect has been found more than once. If the replicate data show the same pattern, it is normal to calculate average results across replicates, presenting these as the final results.

DON'T FORGET

Be critical of your data.

Analysis of your data may show any or all of the following: small sample sizes, large variability in data within samples, lack of clear differences between treatments, or lack of concordance between replicate sets of results. If this is the case, you cannot use your data to conclude that the independent variable has a clear effect on the dependent variable.

Replicate 1

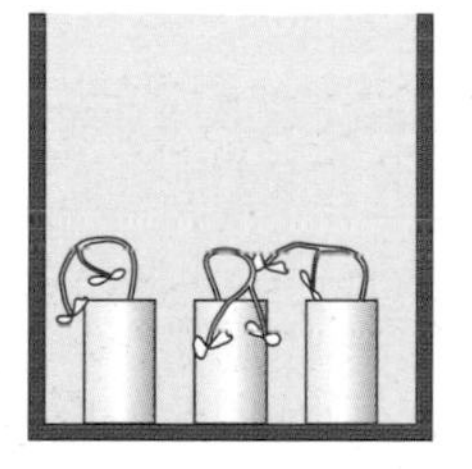

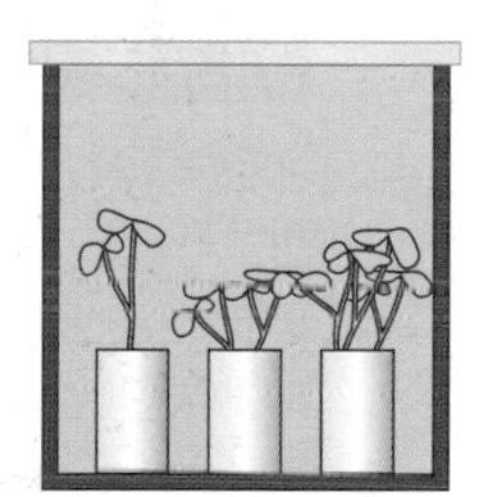

Replicate 2

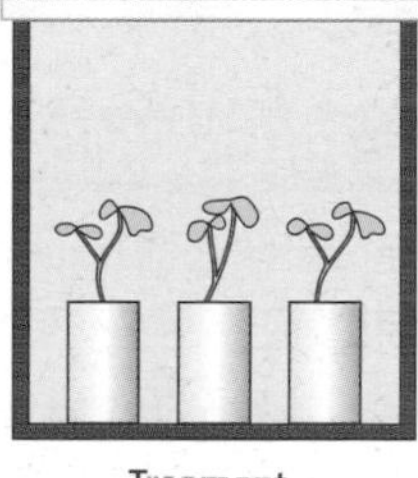

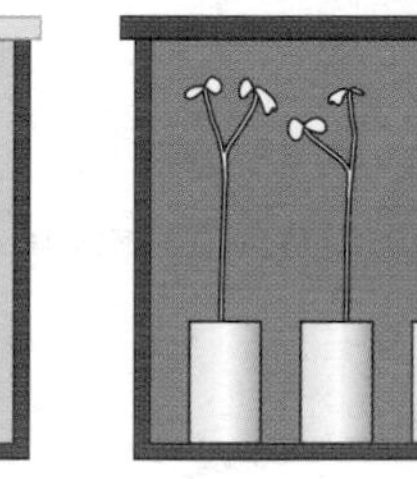

Treament 1 | Treament 2 | Control

An investigation into the effect of light intensity (the independent variable) on seedling growth (the dependent variable). The different treatments of the independent variable are two different light intensities. The control is the absence of light. None of the data from Replicate 1 are valid due to three failures to control different confounding variables. For Replicate 2, the mean results clearly differ for each of the treatments and the control, and the variability in each sample is low. However, a further replicate is required before a conclusion could be drawn. Even then, the very low sample size would make any conclusion very weak.

LET'S THINK ABOUT THIS

Make sure that you know how much time it will take to collect each data point. Only then can you calculate how much time you are planning to spend investigating.

THE INVESTIGATION REPORT

DON'T FORGET

The report must include the following sections:

- title page
- contents page
- summary
- introduction
- procedures
- results
- discussion
- reference list.

MAKE SURE YOU FOLLOW THE INSTRUCTIONS

The Investigation Report submitted to SQA is worth 20% of the final Advanced Higher mark. It is worth ensuring that your report matches the examination requirements as closely as possible; double check using the web reference and the Investigation Mark Allocation (see page 125).

Remember that an examiner will have to mark your work, so make it as **logical** and as easy to read as possible. Communicate **clearly** and honestly, and with enough **detail** so that your investigation could be repeated by anyone who has read your report.

Make sure that all your **raw data** is included; if there is a lot of data, include it as an appendix. The rest of this section offers advice in areas where candidates frequently and unnecessarily lose marks.

Look up www.sqa.org.uk/sqa/files_ccc/NQBiologyAdvHInvestigationGuideMarch08.pdf

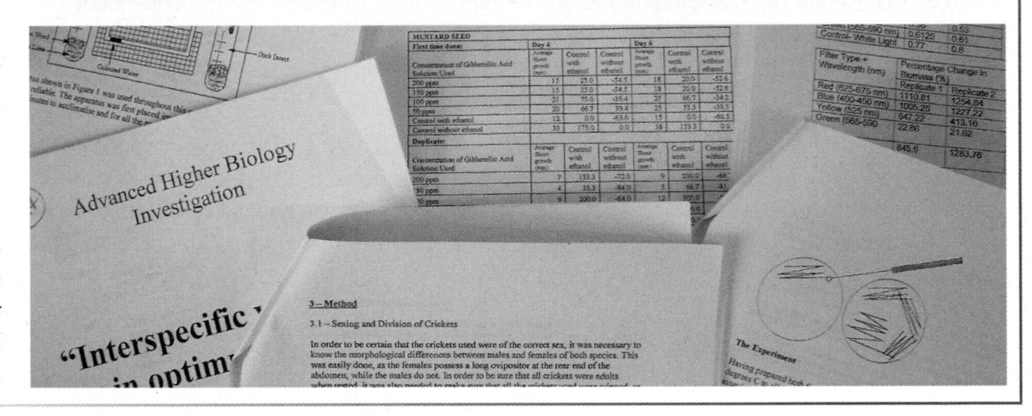

An investigation report should be structured logically and clearly written. The main body of the text should be around 2000 to 2500 words.

YOUR INTRODUCTION

Your introduction must include a clear statement of the aims of the investigation (despite the fact that you have already stated these in the summary), along with relevant hypotheses or questions.

In this section, you must also include an account of the relevant background theory at a level appropriate to Advanced Higher. You are expected to provide plenty of detail. Remember, the examiner may know little about your chosen topic, so your introduction should provide the background information. You must also justify the biological importance of the investigation. Why did you think it was worth doing in terms of biology? Does the topic have wider implications?

GRAPHICAL PRESENTATION

It is essential to summarise your results adequately using graphs. But graphs should not appear alone; every graph that you draw should also have a table showing the relevant processed data.

You may be adept at using electronic software to manipulate the axes of graphs to suit your data, so that the results are presented in a correct scientific manner. If you are at all unsure, you are advised to draw your graphs by hand. Check your graphs to see if they present your data clearly:

- Are the axes labeled and do they include units?
- Is the zero at the start of each axis?
- Are the scales evenly spaced?
- Have you used line graphs for continuous variables and bar graphs for discontinuous variables?
- Have you presented your data sets so that they can be compared easily?

YOUR CONCLUSIONS AND EVALUATION

The discussion section is the most important part of the investigation report. You must refer back to what you have written earlier in the report and discuss your findings in a critical and scientific manner.

In your discussion section, you need to provide a clear statement of the **overall conclusion** stating how your results relate to the biological effect you were investigating. Your conclusion must relate to the aim of the investigation and be valid for the results obtained. Do not be tempted to over-interpret what you have found!

Essentially, this section is your chance to explain whether your independent variable has influenced your dependent variable. First, you should evaluate whether your **procedures** were robust enough for you to be able to detect an effect – or are there issues to do with the accuracy, replication, sample size, controls or errors? Next you should explain very carefully how confident you can be about your **results**. You should discuss the degree of **variation** in your results **within samples, between treatments** and, finally, **between replicates**.

REFERENCES

A **reference** is any piece of material to which a writer **'refers'** in the text. For example, the authors of this book (Lloyd and Morgan, 2010) find that they often refer to Raven and Johnson (2002) and Begon *et al.* (2006) when planning their teaching. The journal *Biological Sciences Review* is also recommended as it contains many articles relevant to Advanced Higher Biology, such as one on growing algae for food (Roberts, 2007). Website references must include the following where possible: author, date, title, publisher, URL and the **date you accessed the material** (because the website may be updated later).

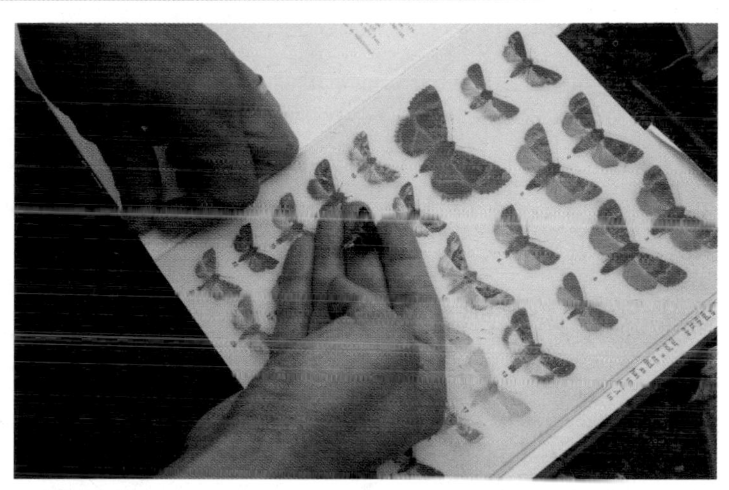

Scientists frequently refer to publications during investigative work.

Example reference list

Begon, M., Townsend, C.R. & Harper, J.L. (2006) *Ecology: From Individuals to Ecosystems (4th Edition)*, Blackwell Publishing, Oxford

Lloyd, D. & Morgan, G. (2010) *Advanced Higher Biology*, BrightRed, Edinburgh

Raven, P.H. & Johnson, G.B. (2002) *Biology (6th Edition)*, McGraw-Hill, New York

Roberts, D. (2007) Aquaculture *Biological Sciences Review* Vol. **20** (No.1), pages 2–6

Scottish Qualifications Authority (2008) *Advanced Higher Biology Candidate Investigation Guidance (for use from session 2008–2009)*, Scottish Qualifications Agency URL: www.sqa.org.uk/sqa/files_ccc/NQBiologyAdvHInvestigationGuideMarch08.pdf (date accessed: December 2009)

DON'T FORGET

You **must** use this style of referencing for Advanced Higher Biology

LET'S THINK ABOUT THIS

When can you be confident that your independent variable affects your dependent variable? Simple: when both the following apply.

1. your procedures are valid, accurate and reliable
2. your results have low variation within samples, low variation between replicates, but significantly different results between your treatments!

EXAM PRACTICE

MULTIPLE CHOICE QUESTIONS

CELL AND MOLECULAR BIOLOGY

1. What is the mitotic index for the following set of cells?

Phase of cycle	Number of cells
Interphase	1120
Prophase	150
Metaphase	90
Anaphase	20
Telophase	20

A 8%
B 20%
C 25%
D 32%

2. One litre of cell culture growth medium containing 15% foetal bovine serum (FBS) can be made using

A 850 ml of medium + 150 ml of FBS
B 985 ml of medium + 15 ml of FBS
C 1000 ml of medium + 15 ml of FBS
D 1000 ml of medium + 150 ml of FBS

3. Which of the following statements about the sodium–potassium pump is correct?

A It results in a higher concentration of sodium ions inside the cell.
B The transport protein has an affinity for sodium ions in the cytoplasm.
C It results in a higher concentration of potassium ions outside the cell.
D The transport protein has an affinity for sodium ions in the extracellular fluid.

4. During DNA replication one of the strands of DNA (the lagging strand) is replicated as a series of fragments that are then bonded together. The enzymes that make and bond the DNA fragments are

A polymerase and ligase
B ligase and kinase
C polymerase and nuclease
D kinase and nuclease.

5. Which line in the table below identifies correctly an extracellular hydrophobic signalling molecule and its action?

	Signalling molecule	Action
A	insulin	activates gene regulatory proteins
B	testosterone	activates receptor proteins on the target cell surface
C	insulin	activates receptor proteins on the target cell surface
D	testosterone	activates gene regulatory proteins

6. The diagram below shows the first two nucleotides of a DNA strand.

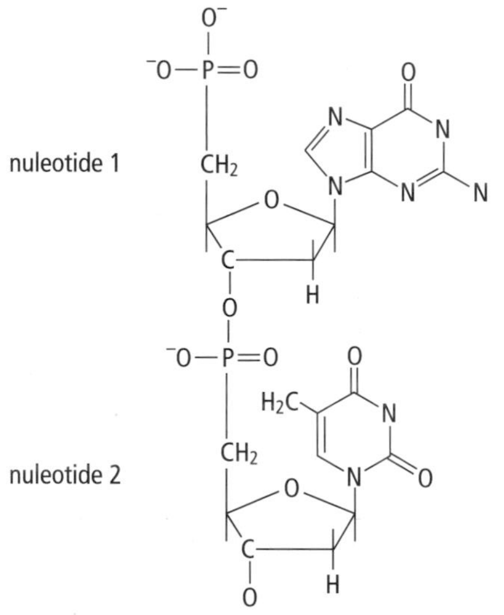

Which of the following statements about the DNA strand is correct? Nucleotide 1 is at the

A 3′ end and has the base guanine
B 3′ end and has the base thymine
C 5′ end and has the base guanine
D 5′ end and has the base thymine.

7. Cytokinins are used in plant tissue culture to

A promote totipotency
B promote differentiation
C produce pathogen-free plants
D fuse protoplasts.

ENVIRONMENTAL BIOLOGY

1. The difference between gross primary production (GPP) and net primary production (NPP) in an ecosystem is due to

   A photosynthesis by autotrophs
   B decomposition by heterotrophs
   C death of heterotrophs
   D respiration of autotrophs.

2. The sea star, *Pisaster ochraceous*, is a key predator of the rocky intertidal zone on the coast of Washington State, USA and it feeds on mussels and other invertebrates. The graph below shows the effect of removing *Pisaster* from a rock pool in 1993.

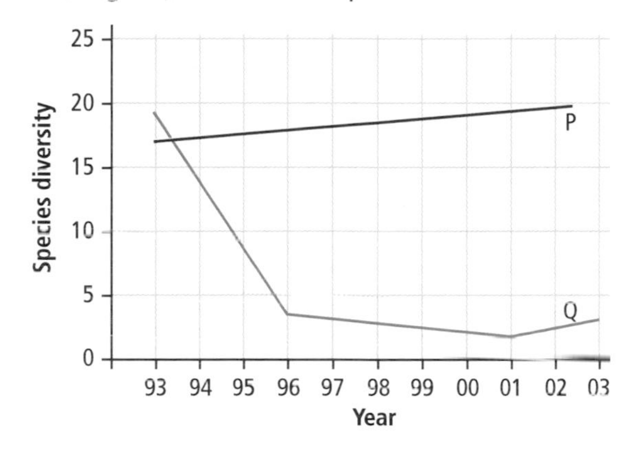

Which line in the table below correctly describes the results?

	Line P	Line Q	Role of *Pisaster*
A	with *Pisaster*	without *Pisaster*	increases species diversity
B	with *Pisaster*	without *Pisaster*	decreases species diversity
C	without *Pisaster*	with *Pisaster*	increases species diversity
D	without *Pisaster*	with *Pisaster*	decreases species diversity

3. Which line in the table may be correctly applied to detritivores?

	Mode of nutrition	Effect on humus production
A	saprotrophic	increased
B	saprotrophic	decreased
C	heterotrophic	increased
D	heterotrophic	decreased

4. The graph below shows the relationship between water temperature and muscle temperature for three species of tuna. Tuna are poikilothermic fish but some have the ability to thermoregulate.

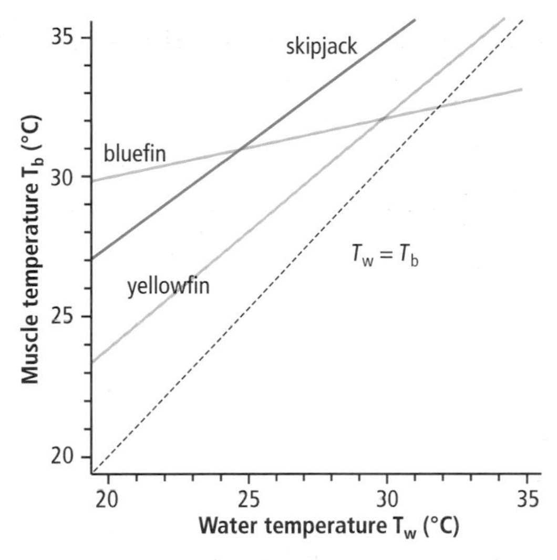

At which environmental temperature would the best thermoregulator have the same muscle temperature as a perfect conformer?

   A 24°C
   B 29°C
   C 30°C
   D 32°C

5. Grazing changes the structure of plant communities as a result of

   A reducing autotroph productivity
   B increasing the biomass of herbivores
   C removing plants with basal meristems
   D removing plants without basal meristems.

6. The widespread distribution of DDT in the environment is a result of

   A toxicity
   B persistence
   C biotransformation
   D biological magnification.

7. The face mite lives in the hair follicles around the eyelids of some humans. It feeds harmlessly on the oily secretions from the sweat glands. Which type of relationship is being described?

   A Competition
   B Commensalism
   C Mutualism
   D Parasitism

SHORT ANSWER QUESTIONS AND ESSAYS

CELL AND MOLECULAR BIOLOGY

1. Sickle cell anaemia is an inherited condition affecting haemoglobin. Four polypeptide chains interact to form a haemoglobin molecule. Two of the chains are designated α chains and the other two are referred to as β chains. Sickle cell anaemia arises from a mutation in the gene for β chains.

The restriction enzyme *Mst*II is able to recognise and cut DNA that has the sequence CCTNAGG, where N is any nucleotide. One of these *recognition sites*, CCTGAGG, lies within the β chain gene. In sickle cell anaemia the mutation has changed the sequence to CCTG<u>T</u>GG. This alteration to the gene can be used to screen for sickle cell anaemia.

a. What type of enzyme is a restriction enzyme? 1

b. Following the extraction of the DNA for the β chain of haemoglobin and its digestion with *Mst*II, what technique is used to separate the fragments produced? 1

c. A probe has been designed to hybridise with the gene for the β chain. Describe the features of the probe that allow it to hybridise. 2

d. i. What is meant a *screening test* for a disorder? 1

ii. Explain why the action of *Mst*II allows it to be used to screen for sickle cell anaemia. 1

2. Describe the organisation of genetic material in prokaryotes and eukaryotes. 4

ENVIRONMENTAL BIOLOGY

1. Species A and species B occupy the same habitat. The figure shows the consumption of prey items of different size by the two species.

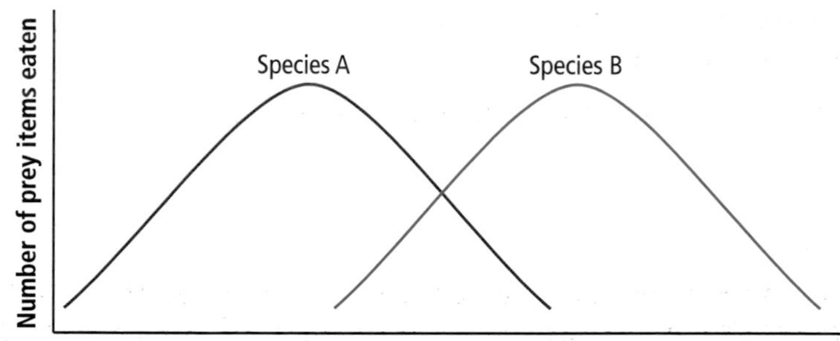

a. Neither species is able to occupy its fundamental niche.
What evidence in the figure supports this? 1

b. Explain how *resource partitioning* would allow the two species to coexist. 1

c. Give one other biotic interaction, not shown on the figure, which could influence the population density of either species. 1

2. Discuss air pollution resulting from human activities under the following headings:

i. use of fossil fuels; 4

ii. the greenhouse effect; 7

iii. impact on abundance and distribution of species. 4

(15)

BIOTECHNOLOGY

1. The diagram shows steps involved in producing monoclonal antibodies.

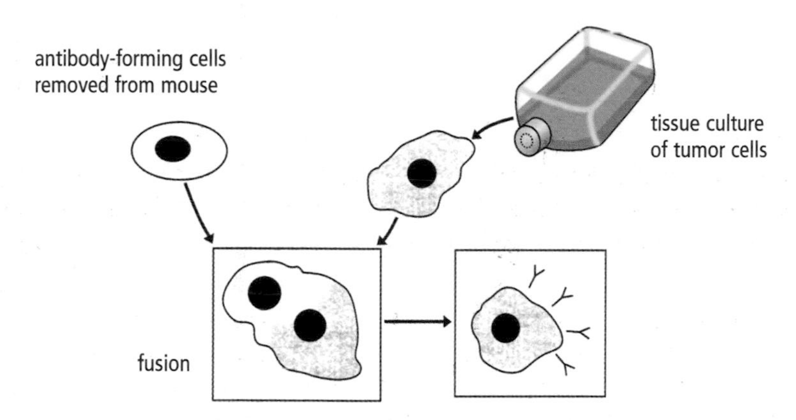

a. Which cell type in the mouse produces antibodies? 1
b. Why are the antibody-forming cells fused with tumour cells? 1
c. What term is used to describe the fused cells? 1
d. The tumour cells selected for fusion have lost their ability to synthesise antibody molecules. Why is this important? 1

2. Discuss the role of enzymes in the commercial production of fruit juices. 4

ANIMAL BEHAVIOUR

1. The capercaillie, *Tetrao urogallus*, is a bird found in some ancient Caledonian pine forests. It is the largest member of the grouse family. Males are larger and noisier than females, particularly in spring when their displays make them very conspicuous.

a. What term is used to describe the condition shown in the capercaillie, where there are distinct differences between males and females? 1
b. State **one** function of the male displays in spring. 1
c. How do females benefit from being inconspicuous? 1

2. Describe the main features of social hierarchies found in primate groups. 5

PHYSIOLOGY

1. Blood lipid profiling is carried out as part of routine health checks to give an indication of the risk of cardiovascular disease. The profile includes measurements of the concentrations of the lipoproteins HDL and LDL.

a. What is the role of lipoprotein in the development of atherosclerosis? 2
b. Describe the effect of regular exercise on blood lipid profiles. 1

2. Outline the factors that bring about variations in total energy expenditure. 5

EXAM PRACTICE

DATA QUESTION

In humans, there are large fluctuations in carbohydrate intake and energy expenditure. The body monitors and responds to changes in the blood glucose concentration. Homeostatic mechanisms ensure a steady concentration of soluble glucose in the blood. In this context, specialised cells in the pancreas secrete the hormones **insulin** and **glucagon**. Excess glucose is stored in the liver as glycogen.

The graphs in the figure show changes in the concentrations of glucose, insulin and glucagon in the blood of a person who had drunk a solution containing 75 grams of glucose. The person had been fasting overnight before drinking the glucose solution. The table shows the effects of insulin and glucagon.

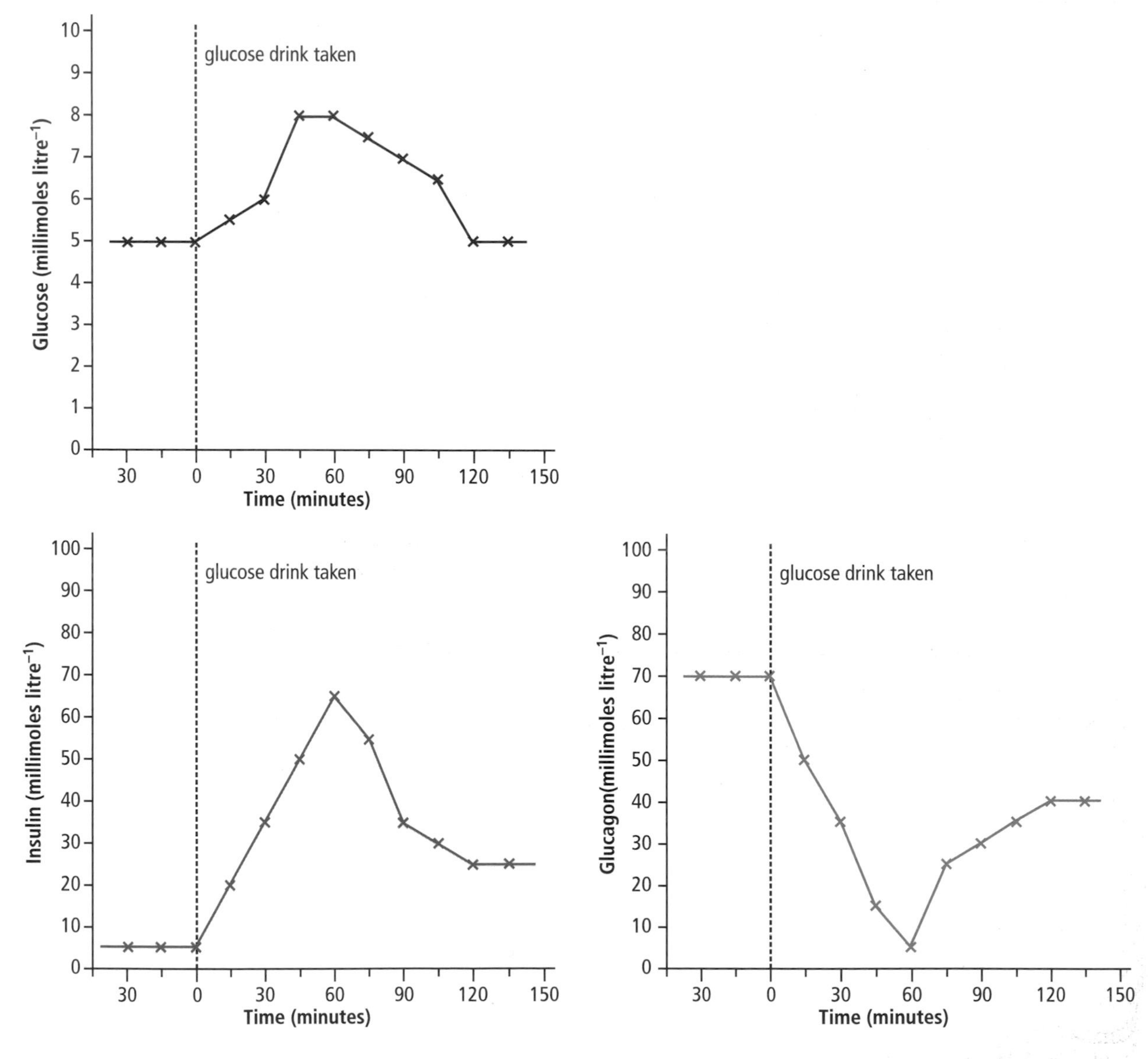

Figure: Changes in concentration of glucose, insulin and glucagon in the blood

contd

Insulin	Glucagon
stimulates the conversion of glucose to glycogen	stimulates the conversion of glycogen to glucose
inhibits the conversion of glycogen to glucose	
inhibits *gluconeogenesis in the liver	stimulates *gluconeogenesis in the liver
increases glucose uptake by muscle cells	decreases glucose uptake by muscle cells
stimulates conversion of glucose to fatty acids	stimulates conversion of fats to fatty acids and glycerol
stimulates fat storage	

***gluconeogenesis** = the production of glucose from lactic acid or glycerol

Table: The effects of insulin and glucagon on metabolism

a. Use the data shown in the figure to describe the changes in blood glucose concentration over the first 60 minutes after the glucose drink was taken. 2

b. i. For an animal cell, what is the advantage of storing glucose molecules in the form of glycogen? 1

ii. Describe the structure of glycogen. 2

c. What ranges of insulin and glucagon concentrations are present in the blood when glucose concentration is 8 millimoles litre^{-1}? 2

d. Insulin is a peptide hormone made from a chain of amino acids.

i. Name the four classes of amino acids. 1

ii. Which type of bond stabilises the secondary structure of insulin? 1

iii. Name the two arrangements of the polypeptide chain that make up the secondary structure. 1

e. During periods of high energy expenditure and low food intake, lactic acid and glycerol are used for gluconeogenesis in the liver. Explain how these two substances have been formed. 2

f. After two days' starvation, the blood glucose concentration falls to 3.5 millimoles litre^{-1}.

i. Predict the changes in the levels of insulin and glycogen. 1

ii. What effects will these changes have on carbohydrate metabolism? 2

(15)

ANSWERS

MULTIPLE CHOICE QUESTIONS

Cell and Molecular Biology

1. B PS AH Biology 2005 Q2
2. A PS AH Biology 2005 Q5
3. B KU AH Biology 2005 Q8
4. A KU AH Biology 2006 Q7
5. D KU AH Biology 2005 Q10
6. C KU AH Biology 2006 Q4
7. B KU AH Biology 2006 Q2

Environmental Biology

1. D KU AH Biology 2005 Q14
2. A PS AH Biology 2005 Q22
3. C KU AH Biology 2006 Q15
4. D PS AH Biology 2006 Q18
5. D KU AH Biology 2005 Q23
6. B KU AH Biology 2005 Q25
7. B KU AH Biology 2005 Q24

SHORT QUESTIONS AND ESSAYS

Cell and Molecular Biology

AH Biology 2006 Q3

1. a. Nuclease/endonuclease 1KU
b. (Gel) electrophoresis 1KU
c. • Base sequence of probe is complementary to DNA base sequence
OR Complementary base sequence
• Binds (to DNA in the extract) by base pairing
• Single stranded Any 2 = 2KU
d. i. Test that distinguishes affected/unaffected/carrier individuals
OR Test picks up if an individual … has/will/might develop a condition 1KU
ii. Cuts/recognises normal site but not the mutated site/sequence/DNA
OR Produces more fragments for normal gene/DNA … than for abnormal gene/DNA 1PS

AH Biology 2005 Q8A (Part)

2. Describe the organisation of genetic material in prokaryotes and eukaryotes.

1. Prokaryotes have single circular DNA molecule (NOT: chromosome)
2. DNA/chromatin/genetic material is condensed/folded/coiled
3. Eukaryotes have nuclear membrane/envelope
4. Prokaryote DNA organised as nucleoid (NOT: 'have no nucleus')
5. Eukaryote DNA organised as chromosomes/associated with histones/associated with proteins
6. Plasmid is an additional ring of DNA/genes

4KU

Environmental Biology

AH Biology 2005 Q3

1. a. Overlap in (range of) food/prey size
OR Sharing/competing for prey of same size 1PS
b. Species A takes smaller food items
OR Species B takes larger food items
OR Reduces competition 1KU
c. Disease
Parasitism Either = 1KU

AH Biology 2004 Q2B

2. Discuss air pollution resulting from human activities under the following headings:

Use of fossil fuels

1. Growing demand for energy from increased population/affluence/standard of living
2. Burning fossil fuels/coal/gas/oil (to obtain energy) …
3. … results in increasing air pollution/emission of gases
4. Two examples from SO_2, NOx, CO_2, CO (Mark not given if mention of CFC or methane)
5. Effect of acidic gases (from burning fossil fuels)

4KU

The greenhouse effect

1. Description of greenhouse effect as allowing incoming radiation but restriction of outgoing radiation/heat (Clear diagrams accepted)
2. Caused by normal carbon dioxide AND water in air
3. Maintains temperature of atmosphere/insulates/retains heat
4. Enhancement of greenhouse effect by increasing addition of pollutants
5. (Water and) carbon dioxide increased by burning/deforestation
6. Greenhouse gases from other sources
7. Example: methane/biogas AND source (e.g. cows, paddy fields)
8. Example: CFC and source (aerosol propellants, refrigerants)
9. (Increased temperature) causes global warming

7KU

Abundance and distribution of species

1. Global warming changes climate/weather
2. Leads to habitat destruction/desert formation/fires, etc.
3. (Relationship between) zooxanthellae and coral
4. (Effect of) increasing sea temperature (from global warming) …
5. … destroys relationship between the two/'coral bleaching'/death of coral
6. Exemplification of how change in environmental conditions could result in change in distribution of species

7. Susceptible species idea where some species will die out in new conditions (e.g. acid rain effects)
8. Tolerant species/favoured species/indicator species idea

4KU

Biotechnology

AH Biology 2006 Q2

1. a. B lymphocytes 1KU

b. To make the cell line immortal
OR So the fused cells can divide indefinitely
OR So they don't need anchorage 1KU

c. Hybridomas 1KU

d. To prevent fused cells producing two kinds of antibodies
OR So that only **one type** of antibody is produced 1KU

AH Biology 2006 Q3

2. Discuss the role of enzymes in the commercial production of fruit juices.

1. Problem: Clarity/haze/cloudiness
2. Solution: Araban–arabinose OR Starch–amylase OR Pectin–pectinase
3. Problem: Filtration/viscosity
4. Solution: Pectin–pectinase
5. Problem: Yield/mechanical extraction/cell wall strength
6. Solution: Cellulose (in cell walls) and cellulose

4KU

Animal behaviour

AH Biology 2005 Q3

1. a. Sexual dimorphism 1KU

b. To attract females
OR Let females select 'best' male 1KU

c. Camouflage decreases chances of predation
OR Improves chances of survival for young 1KU

AH Biology 2006 Q1 (Part)

2. Describe the main features of social hierarchies in primate groups.

1. Hierarchy depends on dominance/system of social ranking
2. (Hierarchy) established by fighting/conflict
OR (Hierarchy) maintained by threat displays/appeasement
3. Rank determines access to food/resources/mates
4. (Hierarchy) reduces actual physical fighting
5. Hierarchies are generally linear
OR Alliances may be formed where animals of subordinate ranks join
6. Rank changes with time as animals grow/reproduce/age
7. Grooming to lower dominance threat
OR Grooming to maintain close relationship/maintain rank
8. Sexual presentation as appeasement gesture

5KU

Physiology, Health and Exercise

AH Biology 2005 Q2

1. a. Lipoproteins transport cholesterol
OR LDL transports cholesterol from liver to cells
OR HDL 'scavenges' cholesterol 1KU
Fatty material deposited in artery walls leads to atheroma/narrowing of arteries 1KU

b. Increases HDL and reduces LDL
OR HDL:LDL ratio increases 1KU

AH Biology 2005 Q4

2. Outline the factors that bring about variations in total energy expenditure.

1. Total energy expenditure affected by basal metabolic rate (BMR) – defined as measure of energy expended to carry out basic body functions
2. (BMR affected by) body size – BMR increases as body mass increases
3. (BMR affected by) body composition – lean tissue more active than fat/adipose
4. (BMR affected by) age – higher BMR in children/lower BMR in adults
5. (BMR affected by) sex – higher in males/lower in females
6. (BMR affected by) nutritional status – reduced by fasting/low energy intake
7. (BMR affected by) thermic activity of food/use of energy to deal with food – different foods affect energy expenditure differently
8. (BMR affected by) physical activity – energy expenditure above resting/more activity gives higher energy expenditure

5KU

DATA QUESTION

Modified from CSYS Biology 1995

a. *Must have full quantification of time and concentration with units used at least once*

0–30 minutes – Glucose concentration rises from 5 to 6 millimoles litre^{-1}

30–45 minutes – Glucose concentration rises from 6 to 8 millimoles litre^{-1}

45–60 minutes – Glucose concentration stays at 8 millimoles litre^{-1}

Any 2 = 1, All 3 = 2PS

(If all times given but no glucose units = 1 mark)

(If no times given but all glucose concentrations are correct = 1 mark)

b. i. Free glucose is soluble while glycogen/ polysaccharide is not

If carbohydrate was stored as glucose, then water would be drawn into the cell by osmosis

Both = 1KU

ii. α-glucose (monomer)

α-1,4 chain

α-1,6 branches

Any two = 1, All 3 = 2KU

c. Insulin: 50–65 millimoles litre^{-1} 1PS

Glucagon: 5–15 millimoles litre^{-1} 1PS

Units must be given to gain the marks.

(If numbers are correct but units not given both times = 1 mark)

d. i. Acidic, basic, polar, non-polar 1KU

ii. Hydrogen bonds 1KU

iii. α-helix AND ß-sheet Both = 1KU

e. High energy expenditure leads to anaerobic respiration so lactic acid is produced 1PS

Low food intake, glucagon levels are high so stimulate conversion of fats to fatty acids and glycerol 1PS

f. i. Low insulin and high glucagon 1PS

ii. Glycogen to glucose

Gluconeogenesis converts glycerol to glucose

Reduced uptake of glucose by muscles

Any 2 = 1, All 3 = 2PS

INVESTIGATION MARK ALLOCATION

Use this table carefully to check your investigation report.

Assessment category and criteria	*Mark*	*Done*
Presentation (3 marks)		
• Appropriate and informative title		
• Contents page and page numbers throughout the report		
• Brief summary stating aims **and** findings	1	
• Three references from different sources cited in text and listed in standard format (two must be books or journals)	1	
• Report is clear and concise (2000–2500 words)	1	
Introduction (4 marks)		
• Clear statement of Aims, together with Hypotheses	1	
• Account of underlying biology relevant to Aims		
• Biological ideas are clear and at an appropriate depth for Advanced Higher		
• The biological importance of the investigation is explained and justified	3	
Procedures (6 marks)		
• Methods used are valid and appropriate to the Aims	1	
• Clear description of methods with enough depth to allow repetition by others	1	
• Appropriate controls and adequate control of confounding variables	1	
• Sample size large enough and whole experiment replicated	1	
• Methods are of appropriate complexity for Advanced Higher		
• The investigation involves creativity and originality		
• Methods are of appropriate accuracy or modified to improve accuracy	2	
Results (5 marks)		
• Results are relevant to the original Aims	1	
• The data is recorded within the levels of accuracy of measurement (there are no excess decimal places in raw or processed data)	1	
• The data presented summarise the overall results	1	
• High quality of tables (headings, units) and graphs (scales, units, labels, clarity)	1	
• Description of trends and patterns shown by tables or graphs	1	
Discussion (7 marks)		
• The conclusions relate to the Aims of the investigation	1	
• The conclusions are supported by the results obtained	1	
Evaluation of procedures includes comments, as appropriate, on:		
• Accuracy/sources of error in measurements		
• Adequacy of replication/sampling		
• Adequacy of controls		
• Solutions to problems and modifications to procedures	2	
Evaluation of results includes comments, as appropriate, on:		
• Analysis and interpretation of results		
• Account taken of error/variation in replicates		
• Meaningful suggestions for further work		
• Critical and scientific discussion of significance of findings		
• Appropriate depth of biological knowledge and understanding	3	
Total marks	25	

INDEX